Knowledge Media in Healthcare: Opportunities and Challenges

Rolf Grütter
University of St. Gallen, Switzerland

Idea Group
Publishing

Information Science
Publishing

Hershey • London • Melbourne • Singapore • Beijing

Acquisition Editor: Mehdi Khosrowpour
Managing Editor: Jan Travers
Development Editor: Michele Rossi
Copy Editor: Maria Boyer
Typesetter: LeAnn Whitcomb
Cover Design: Deb Andree
Printed at: Integrated Book Technology

Published in the United States of America by
        Idea Group Publishing
        1331 E. Chocolate Avenue
        Hershey PA 17033-1117
        Tel: 717-533-8845
        Fax: 717-533-8661
        E-mail: cust@idea-group.com
        Web site: http://www.idea-group.com

and in the United Kingdom by
        Idea Group Publishing
        3 Henrietta Street
        Covent Garden
        London WC2E 8LU
        Tel: 44 20 7240 0856
        Fax: 44 20 7379 3313
        Web site: http://www.eurospan.co.uk

Library of Congress Cataloging-in-Publication Data

Grütter, Rolf, 1958-
   Knowledge media in healthcare : opportunities and challenges / Rolf Grütter
      p. cm.
   Includes index.
   ISBN 1-930708-13-0
   1. Medical informatics. 2. Health services administration--Data processing. I.Title.

R858..G78 2001
362.1'068'4--dc21   2001024743

British Cataloguing in Publication Data
A Cataloguing in Publication record for this book is available from the British Library.

Knowledge Media in Healthcare: Opportunities and Challenges

Table of Contents

Section V: Towards a Global Knowledge Medium

Preface

INTRODUCTION TO THE BOOK

Healthcare has always been both an opportunity and challenge for emerging technologies. For instance, the discovery of X-rays by Wilhelm C. R. Röntgen in 1895 created the opportunity for diagnostic imaging techniques to be developed in the 20th century. Specifically, it provided the basis for X-ray testing. Coupled with the mathematical foundation introduced by Johann Radon in 1917, the so-called Radontransformation, and digitization, it further enabled the invention of the computer tomography scanner by Godfrey N. Hounsfield and Allan M. Cormak in 1972 (Hounsfield and Cormak were awarded the Nobel award for medicine in 1979 for their invention). The implicit challenge lies in the long lag between discovery and invention.

Similarly, the incremental and still ongoing diffusion of information technology (IT) into healthcare has markedly changed the environment in which healthcare is studied, taught, and practiced. In addition, it has energized the establishment of the novel discipline of medical informatics. *Medical informatics* is evolving in two main directions. These are signal processing and information management. *Signal processing* in healthcare is applied, e.g., in equipment. *Equipment* is comprised of sensors and motors. For instance, *sensors* are used for monitoring the state of a patient (circulation, respiration, chemicals, etc.) during anesthesia or under emergency condition, while *motors* are applied in computer-supported (i.e., micro-) surgery.

Information management in healthcare is understood in a broad sense and includes a rich variety of activities. To mention a few: documentation of the patient's condition, diagnostic measures, treatment, etc.; monitoring of the patient's compliance during long-term treatment; diagnostic coding and other terminology-based application; clinical data warehousing; case-based reasoning; expert systems for drug prescription, second opinion, and education; information services for evidence-based medicine; and tools supporting the transfer and management of medical knowledge in practice.

With the advent of the Internet as a global communications infrastructure and of the World Wide Web as an interactive, ubiquitous, and multimedia information service the opportunities and challenges of IT in healthcare have gained a new dimension. It is within this context of an evolving "global knowledge medium" that the topics presented in this book must be considered. With respect to the rough classification, as introduced above, this book refers to the broad field of information management. The research question, which implicitly underlies this book's concept, can be sketched as follows: *How will the recent trend towards knowledge media change the way things are done in healthcare?* Because of the nature of the question a definite answer cannot be given at this time. However, the contributions presented here, coming as they do from the frontlines of research into IT and its application to healthcare, provide reasonable evidence on how healthcare *might* evolve in the future. Interestingly, the contributions to the book originate from quite different domains, i.e., computer science, medical science, and business sciences. This reflects the requirement of a multidisciplinary approach to the topic that is needed, in order to meet its challenges effectively.

CONTENTS OF THE BOOK
Foundations of Knowledge Media

Stanoevska-Slabeva provides a definition of knowledge media and an over-view of their past and future. The term "knowledge media" was coined and their initial vision introduced by Mark J. Stefik. Stefik defines knowledge media "as an information network with semi-automated services for the generation, distribution, and consumption of knowledge." The first attempts to implement Stefik's concept of knowledge media were conducted at Stanford University by the Knowledge Sharing Effort of DARPA. The result of the joint effort is the concept for an Agent Communication Language (ACL), which enables interoperability and knowledge reuse among heterogeneous artificial agents. ACL consists of an inner language called Knowledge Interchange Format (KIF), an outer language called Knowledge Query and Manipulation Language (KQML), and common vocabularies, i.e., ontologies. Particularly, *ontologies* are considered the basic elements of knowledge media. *Stanoevska-Slabeva* then presents the concept of knowledge media as introduced by Beat F. Schmid from the University of St. Gallen. While adhering to Stefik's vision and the Stanford approach, it extends the original conception in several ways, particularly by introducing the concept of the *community*. The approaches of the Knowledge Media Design Institute at the University of Toronto and of the Knowledge Media Institute at the Open University in London are then presented. Even though these are not directly grounded in the original vision they reflect basic features of knowledge media. Finally, an overview of the Internet as the future global knowledge medium is provided. In this context, the recent initiative towards a *Semantic Web* is discussed.

Lanzola and Boley report on their experience with applying a functional-logic language to distributed, multi-agent medical problem solving. RELFUN, used both as the agent-implementation and the communication-content language, cross-extends Horn relations and call-by-value functions just enough to yield a unified operator concept. Particularly, relations acquire application nesting and higher-order notation; functions acquire non-groundness and non-determinism. Relations are defined by Horn-like clauses; functions are defined by rules with an additional returned-value premise. The chosen multi-agent framework relies on the communication protocol of KQML. The selected medical problem concerns task planning for patients affected by AML (acute myeloid leukemia). The medical professionals involved in AML treatment are modeled as agents communicating in a well-defined fashion. On the basis of a RELFUN representation of KQML performatives, both cooperation and domain knowledge are formalized in a functional-logic style. Its declarative power is shown to be useful for this application. However, for communicating clausal knowledge, update operations turn out to be unavoidable, and RELFUN is extended towards assert/retract. The efficiency of the interpreter has been sufficient in this communication-intensive prototype.

Boley examines mutual relationships between logic programming and XML. XML documents are introduced as linearized derivation trees and mapped to Prolog structures. Conversely, a representation of Herbrand terms and Horn clauses in XML leads to a pure XML-based Prolog (XmlLog). The XML elements employed for this are complemented by uses of XML attributes like id/idref for extended logics. XML's document type definitions are introduced on the basis of ELEMENT declarations for XmlLog and ATTLIST declarations for id/idref. Finally, queries in languages like XQL are treated functional-logically and inferences on the basis of XmlLog are presented. All concepts are explained via knowledge-representation examples from the discourse domain *E-health*.

Knowledge Representation and Human-Computer Interaction
Straub analyzes four different types of classification models. Classification models are intensely used in healthcare, e.g., for the coding of diagnoses or indexing of medical subjects. Starting from the problem of mapping clinical facts onto an artificial representation he introduces step by step more complex models, thereby evaluating each of them against the real world. With the fourth generation of classification models, i.e., the *multifocal, multi-point model*, which approximates real world best, he surpasses existing research and makes an original contribution to the topic. The presentation is supported by a rich number of examples from the practice of diagnostic coding.

Wagner argues that pharmaceutical expert information is characterized by steady increase in extent and complexity as well as by strong interweavement and penetration with interdependent references. These properties suggest the utilization of modern hypermedia technology, organizing knowledge as a network of linked pieces and enabling more intuitive, faster and easier navigation within the growing information space. He concludes that there is a need for a suitable, role-oriented *external* representation of pharmaceutical expert information. *Wagner* further argues that the application of pharmaceutical expert information when analyzing medications for contraindications and interactions is mainly mechanical in nature. It requires the ability to follow references and compare codes. The electronic availability of medication information, which is often forced by drug dispensing systems, is not exhausted when these procedures are performed manually at the screen. He concludes that there is a need for automation of these procedures. This automation requires a suitable *internal* representation of pharmaceutical expert information. *Wagner* further argues that the universality of pharmaceutical information systems leaves open their conceptual environment. The different kinds of application scenarios pose strong requirements on the formalization of expert information. He concludes that a *universal representation* has to be found which captures the nature of pharmaceuticals including common properties of drug classification systems. The diversity of the latter within the different healthcare organizations is to be regarded as the most serious barrier aggravating global communication within the healthcare system.

Kapetanios presents a semantically advanced query language for medical knowledge and decision support which enables a system-guided construction (formulation) of meaningful queries based on *meaning* of terms as they appear in medical application discourses. Representation of meaning and/or semantics is achieved by the structure of a knowledge space, i.e., a constraints-based graph formalism. This knowledge space is consulted by an inference engine which drives the query construction process. Both the knowledge space and the inference engine underlie the specification of MDDQL as the *Meaning Driven Data Query Language* at the core of the system. MDDQL requires neither learning of a particular language-based syntax for querying a data repository nor understanding of the underlying database schema in terms of adequate interpretations of data constructs. It embeds natural language based interpretation of the meaning of acronyms used as attributes and/or values and considers the context which relates to the data as expressed in terms of measurement units, definitions, constraints, etc. The system has been and is currently in use in various cardiological and gynecological hospitals and clinics in Switzerland.

Secure Healthcare Application and Data
Knorr and Röhrig argue that even though security requirements in healthcare are traditionally high, most computerized healthcare applications lack sophisticated security measures or focus only on single security objectives. Their chapter

describes special security problems that arise when processing healthcare data using public networks such as the Internet. *Knorr and Röhrig* propose a structured approach using a context-dependent access control mechanism over the Internet as well as other security mechanisms to counter the threats against the major security objectives: *confidentiality, integrity, availability, and accountability*. The feasibility of the proposed security measures is shown through a prototype, which has been developed in a research project focussing on security in healthcare.

The concept of *context-dependent access control* has emerged during the last years: Information about the state of a process model of a working environment is combined with general knowledge about a person to grant or revoke access to protected data. Being understood very well in principle, different problems arise when implementing context-dependent access control, in particular on an open network. *Ultes-Nitsche and Teufel* report on an ongoing project on context-dependent access control to support distributed clinical trials. Centrally stored data will be accessed from contributors to the clinical trial over the Internet. They present in their chapter how context-dependent access control can be implemented on the Internet in a secure way. Technically they use Java servlets to implement the access control and SSL to secure communication. The whole framework is built around the Java Web server. *Ultes-Nitsche and Teufel* emphasize the technical aspects of this scenario in their chapter.

Knowledge Transfer and Management

Shaughnessy, Slawson and Fischer argue that traditional university-based teaching methods equip students with vast amounts of information but they fail to teach the skills for continuous learning. While computers and the Internet have made information readily available at everyone's fingertips, little consideration has been given to how information is delivered. The primary focus of evidence-based medicine is to identify and validate written information; for most doctors this is a too time-consuming process. They conclude that if best available evidence is to be used at the point of care, sophisticated *filters* are needed that increase the yield on relevance of the information. Doctors need an *alert method* for becoming aware of relevant new information that implies the need to update their knowledge; these systems should be tailored to the doctor's individual needs.

Similarly, *Khan, Bachmann, and Steurer* argue that with the enormous, rapid and exponential expansion in the medical literature, there is a need for effective and efficient strategies to keep abreast of relevant new knowledge. The *medical journal club*, traditionally used as an educational tool in postgraduate and continuing medical education, can be designed for acquisition and appraisal of relevant, current best clinical evidence by using a systematic approach to both acquisition and appraisal of evidence in a context directly related to patient care. According to *Khan, Bachmann, and Steurer*, incorporation of computer technology in the journal club helps with acquisition, appraisal and refinement of knowledge, but more importantly it allows for knowledge transfer by making possible the storage and instant retrieval of appraised topics in the future.

As mentioned, there is an enormous amount of new medical knowledge generated every year as a result of research. However to make a practical difference, this knowledge needs to be disseminated and used in everyday medical practice. *Moody and Shanks* describe a knowledge management project which provides medical staff with on-line access to the latest medical research at the point of care, in order to improve the quality of clinical decision making and to support evidence-based practice. The project has been highly successful, and a survey of medical staff using the system found that over 90% felt that it had improved the quality of patient care. They describe how the system was developed and implemented, its functional characteristics and organizational impact. The theoretical significance of their work is that it is one of the first empirical studies of a knowledge management project in the public sector. The practical significance of their research is that it provides a model for other similar organizations to follow in implementing such a project. Finally, *Moody and Shanks* draw some wider lessons from this case study for the practice of healthcare.

Towards a Global Knowledge Medium
According to *Fierz*, medical information processing by machines depends on the accessibility of medical information in electronic form. The manner in which such information is expressed and perceived, both by computers and by human beings, is fundamental to the success of any attempt to profit from modern IT in healthcare. He argues that medical information resides not, to a large extent, in the data content of information itself but is rather distributed within the connections (links), as well as within the structured context of the data elements. To enable a computerized processing of such information, some basic requirements for a structured and linked data model have to be fulfilled. These include the *granularity* of data elements; the way to attach *semantic information* to the data elements, links and structures; the *storage* of the data together with their structure and the connections between the data elements; a *query system* for the extraction of the information contained within the structure and connectivity as well as from the data proper; and the *display* of the query result in a way that structure and connectivity are intuitively and usefully expressed and can be stored again in a structured, machine-accessible way. *Fierz* argues that the

developing Internet technology provides a suitable model for how these requirements can be fulfilled. The *hypertext* paradigm together with the *markup* technology for structuring information provides all the necessary ingredients for developing information networks that might be called *Medical Data Webs*. On top of that, Topic Maps and the Resource Description Framework (RDF) can be employed to semantically navigate through these Medical Data Webs.

RDF(S) constitutes a newly emerging standard for metadata that is about to turn the World Wide Web into a machine-understandable knowledge base. It is an XML application that allows for the denotation of facts and schemata in a Web-compatible format, building on an elaborate object-model for describing concepts and relations. Thus, it might turn up as a natural choice for a widely useable ontology description language. However, its lack of capabilities for describing the semantics of concepts and relations beyond those provided by inheritance mechanisms makes it a rather weak language for even the most austere knowledge-based system. *Staab, Erdmann, Maedche, and Decker* present an approach for modeling ontologies in RDF(S) that also considers axioms as objects that are describable in RDF(S). Thus, they provide flexible, extensible, and adequate means for accessing and exchanging axioms in RDF(S). Their approach follows the spirit of the World Wide Web, as they do not assume a global axiom specification language that is too intractable for one purpose and too weak for the next, but rather a methodology that allows (communities of) users to specify what axioms are interesting in their domain.

With the objective to facilitate integration of information from distributed and heterogeneous sources *Grütter, Eikemeier, and Steurer* describe the integration of an existing Web-based ontology on evidence-based medicine into the RDF/RDFS framework. Their contribution rests upon the application of methods which are to some extent already in place to a real-world scenario. Based on this application, the scope of the term "ontology" within the RDF/RDFS framework is redefined, particularly by introducing a *Simple Ontology Definition Language* (SOntoDL). This redefinition contributes to the implementation of the Semantic Web and to ontology modeling in general.

CONTRIBUTION OF THE BOOK

Knowledge Media in Healthcare: Opportunities and Challenges identifies areas of current research into knowledge media in healthcare on the basis of relevant examples from the fields of knowledge engineering, knowledge representation, human-computer interaction, application and data security, knowledge transfer, and knowledge management. Results from this research show *opportunities* for improved management of medical data, information, and knowledge, thereby supporting the provision of high quality and cost-efficient patient care.

The *challenges* rest upon the degree to which these opportunities are exploited, the ease with which technological innovations are made available, transferred into, and adopted from practice. Possible obstacles thereby include a mismatch between applications and users'–i.e., healthcare professionals'– needs, technology-adverse behavior, and technological illiteracy. In order to anticipate the first two, a lot of investigation into human needs is still required. Such investigations must not only focus on information needs but also on basic needs and how they can be met in an increasingly changing environment. Technology must never hinder but must further the development of human personality and social competency. In order to overcome technological illiteracy, investment into education, at all levels, is necessary. Hopefully, this book can contribute to this endeavor.

ACKNOWLEDGMENTS
It is particularly due to the significant research into knowledge media by Professor Dr. Beat F. Schmid, head of the Institute for Media and Communications Management, University of St. Gallen, that the editor was inspired to adopt the generic concept of knowledge media to the herein presented application domain. The editor wishes to thank the authors for their valuable insights and excellent contributions to this book. Likewise, the help of those involved in the review process, without whose support the project could not have been satisfactorily completed, is gratefully acknowledged. A further special note of thanks goes to the staff at Idea Group Publishing, who provided professional support throughout the whole process from inception of the initial idea to final publication.

Rolf Grütter, DVM, MSc
St. Gallen, Switzerland
March 2001

Chapter I

The Concept of Knowledge Media: The Past and Future

Katarina Stanoevska-Slabeva
University of St. Gallen, Switzerland

INTRODUCTION

Knowledge is the internal state of humans that results from the input and processing of information during learning and performing tasks. According to Nonaka (1991), we can distinguish two kinds of knowledge: tacit and explicit knowledge. Tacit knowledge is highly personal and is deeply rooted in an individual's actions and experience as well as in his ideas, values and emotions. This type of knowledge is difficult to formalize, to communicate, and to share. Explicit knowledge can be expressed independently from its human carrier in the form of data, scientific formulae, specifications, manuals, experience, project reports, and the similar. In its externalized form, "Knowledge is information that changes something or somebody–either by becoming grounds for action, or by making an individual (or an institution) capable of different or more effective action" (Drucker, 1991). As a result, knowledge is considered the most valuable resource in the information age.

Knowledge acquisition, i.e., its transformation from tacit to explicit knowledge, as well as knowledge storage and sharing are two major concerns of research into knowledge management. How can we capture tacit and explicit knowledge held by humans and organizations, codify, store and make it available for further use in an independent manner from its human creator?

Throughout history, several media have appeared, which are capable of storing and transporting knowledge. They can be classified in two groups (see also Armour, 2000):

- natural or human bound media, such as DNA and the human brain, and
- media independent of humans, such as paper and digital media.

The media differ with respect to knowledge persistency, update speed, intentionality, reach and "activeness," i.e., ability to change the outside world (Armour, 2000). The most recent media for knowledge storage and transportation are digital media resulting from the convergence of information and communication technology. When compared to other media, for example, paper, digital media have had unknown features: They are interactive, ubiquitous and multimedia. Digital media can also store procedural knowledge as algorithms (see also Armour, 2000; Stefik, 1986) and can, to a certain extent, mimic human intelligence. In addition, they are ubiquitous and connected to a worldwide network. They make the knowledge they carry ubiquitously available, active and interactive. These new digital media have revolutionized knowledge management and enabled it in a manner not known before. In order to denote these new media the term knowledge media is used. But what exactly are knowledge media?

In this chapter, the history of the concept and different approaches for the implementation of knowledge media as well as its future will be discussed. In section 2, the term knowledge media will be defined. In section 3, the original concept for knowledge media as defined by Stefik (1986) will be described. Section 4 describes current approaches and section 5 elaborates on the Internet as the future knowledge medium. Section 6 provides a summary of the chapter.

MEDIA AND KNOWLEDGE MEDIA– A DEFINITION

To increase the understandability of the concept of knowledge media, first the basic terms related to it will be explained.

We define the term *knowledge* as the internal state of human beings that results from the input and the processing of information. This definition implies that knowledge, in a narrow sense, must be associated with human beings (Schmid, 1997b). Before knowledge can be shared, it has to be externalized on an external carrier (Nonaka, 1991). The basic means for knowledge exchange is communication. As a result, knowledge exchange takes place in communication spaces, which Nonaka and Konno (1999) call "Ba." A "Ba" can be a physical meeting space or a virtual space created by digital media.

Knowledge management refers to all management activities necessary for effective creation, capturing, sharing, and managing of knowledge (Probst, Raub & Rohmhard, 1997). Knowledge management is substantially enhanced by information and communication technology.

Media is the plural for the English word medium. The word medium stems etymologically from the Latin word *medius,* meaning the middle, that which lies in the middle or an intervening body or quantity (see Schmid et al., 1999). A further definition is given in the on-line version of The American Heritage Dictionary (2000), according to which a medium is an intervening substance through which something else is transmitted or carried.

Each science has a special usage for the word medium, so that we can observe different meanings around the basic meaning of it as the middle. Here some examples:

- According to the Brockhaus Encyclopedia, media are the communication means for dissemination of knowledge by way of graphics, signs, print or digital and broadcast media (Schmid et al., 1999).
- In the science of communications, media are defined as the physical or technical means of converting a communication message into information capable of being transmitted along a given channel (Hill and Watson, 1997; Schmid et al., 1999).
- In the context of computer science media are defined as "objects on which data can be stored" (The Online Encyclopedia, 2000).

The most impressive and most cited source of insights into the word media is provided by McLuhan (1964). He defines media as "extensions of ourselves" and writes in his book *Understanding Media*:

....***The medium is the message***. *This is merely to say that the personal and social consequence of any medium–that is, of any extension of ourselves–result from the new scale that is introduced into our affairs by each extension of ourselves, or by any new technology.*

All media are active metaphors in their power to translate experience into new forms.

Against the background of the above definitions, we define knowledge media as extensions of ourselves capable of storing and transmitting explicit knowledge over space and time. Given the different types of media, we can distinguish a broad and a narrow meaning of the word knowledge media. In a broad sense the word knowledge media refers to any medium capable of storing knowledge, i.e., it refers to human and external knowledge media such as paper and digital media (Armour, 2000).

In the narrow sense, knowledge media means the knowledge exchange spaces arising around digital media. Given the interactive, functional, and multimedia character of digital media they are not only extensions of ourselves but additionally create virtual spaces ("Ba"), providing a communication means on which knowledge can be used, shared, and communicated. In this chapter the term knowledge media will be used in its narrow meaning.

THE ORIGIN OF THE KNOWLEDGE MEDIA CONCEPT

Although first origins of the concept of knowledge media appear in writings of Bush, Licklider, Engelbart, Nelson and Kay (see Baecker, 1997), the first comprehensive vision of knowledge media based on digital media was introduced by Stefik in 1986. Stefik (1986) defines knowledge media "*as an information network with semi-automated services for the generation, distribution, and consumption of knowledge.*" In terms of knowledge processes, a knowledge medium is characterized as "*the generation, distribution, and application of knowledge, and secondarily in terms of specialized services such as consultation and knowledge integration*" (Stefik, 1986). According to Stefik's vision, in a knowledge medium, integrators would combine knowledge from different sources, translators would convert different jargons and languages, and summarizers and teachers would enable the usage of knowledge.

Knowledge media can have different shapes and components. Stefik proposed a whole comprised of basic components, isolated and specialized knowledge bases, and expert systems as well as connecting communication networks. Knowledge media can be built by interconnecting isolated knowledge bases or they can even arise from nearby media, for example, communication media, which are not connected to knowledge services and databases.

According to Stefik (1986), the goal of building knowledge media is to tie expert systems and communication media together into a greater whole. He furthermore pointed to the following research questions that had to be answered as a necessary prerequisite for being able to build knowledge media (Stefik, 1986):

1. development of common vocabularies, i.e., terminologies for important domains, and translators, which, based on common vocabularies, are capable of translating concepts from one expert system to another and for combining knowledge from different sources;
2. development of transmission languages in addition to representation languages, which will enable sharing, transportation, and integration of knowledge from different sources;
3. development of knowledge markets with market coordination mechanisms for distribution and renting of knowledge.

In summary, Stefik envisioned interoperating expert systems, i.e., networks of artificial intelligence sources, which communicate by common knowledge representation and communication languages. The combination of communication media with interoperating expert systems results in a new "intelligent" knowledge and communication infrastructure. On the one hand, this infrastructure might be applied for the solution of complex problems requiring combined knowledge from several expert areas and on the other hand might enable communication without language barriers as well as fast sharing and diffusion of knowledge.

In his visionary article, Stefik compares the potential effect of such a new information and communication infrastructure with the impact that resulted from past introductions of new, large-scale standardized transportation and communication infrastructures, for example, roads and railroads. Throughout history, the introduction of a new infrastructure resulted in fast modernization and revolutionized the flow and diffusion of knowledge (Stefik, 1986).

In 1999 Stefik reflected again on his initial concept for knowledge media in his book *The Internet Edge* (Stefik, 1999). Most of his ideas remain little changed and refer to an intelligent infrastructure and the application of artificial intelligence concepts to achieve it. One important extension is the addition of documents to expert systems as building blocks for knowledge media.

RESEARCH APPROACHES FOR THE IMPLEMENTATION OF KNOWLEDGE MEDIA

The initial vision of knowledge media presented by Stefik, is used at various universities working on the implementation of knowledge media. During implementation, the initial definition for knowledge media, as given by Stefik, has been broadened and enhanced. The most interesting approaches directly related to Stefik's concept are:

- the developments at Stanford University and
- the developments at the Institute for Media and Communications Management at the University of St. Gallen.

In addition to the above two approaches, there are other initiatives at other universities that go under the heading of knowledge media, but which are not directly related to the concept of Stefik. The best known examples of other knowledge media concepts are:

- the research at the Knowledge Media Design Institute and
- the research at the Knowledge Media Institute.

All four of these respective concepts will be described in more detail in the following sections.

Development of Knowledge Media at Stanford University

The first attempts to implement the concept of knowledge media, as proposed by Stefik, were conducted at Stanford University by the Knowledge Sharing Effort of DARPA around 1990. The aim of the Knowledge Sharing Effort was the development of techniques, methodologies, and software tools for knowledge sharing and knowledge reuse. The Knowledge Sharing Effort is organized around a consortium consisting of four research labs at Stanford University: the Center for Information Technology (CIT), the Logic Group, the Knowledge Systems Laboratory, and the Database Group. The results of the joint efforts are the concepts for an Agent Communication Language (ACL), which enables interoperability and knowledge reuse among heterogeneous artificial agents. The proposed ACL consists of three parts (Genesereth & Ketchpel, 1994):

- An inner language called Knowledge Interchange Format (KIF).
- An outer language called Knowledge Query and Manipulation Language (KQML).
- Common vocabularies, i.e., ontologies (Gruber, 1993).

KIF is an interlingua represented in first-order predicate calculus, which can be used as a generic representation formalism for the expression of the internal knowledge base of an agent. In the case that an artificial agent applies a specific representational formalism for knowledge representation it can translate it in KIF and communicate it to other agents (Genesereth & Ketchpel, 1994; Wagner, 2000). Thus, *"KIF can be used to support the translation from one content language to another or as a common content language between two agents which use different native representation languages."* (Finin, Fritzson, McKay & McEntire, 1994).

In addition to a common representation of knowledge, communicating agents need a common communication language, as it refers to mutual requests and answers. This role is taken over by the Knowledge Query and Manipulation Language (*KQML*) (Labrou & Finin, 1997; Finin, Labrou & Mayfield, 1997). KQML is a general purpose, high-level, message-oriented communication language, i.e., a transmission language, enabling interoperability and the communication of software agents. KQML is a protocol for information exchange, independent of content syntax and ontology. Each KQML message represents a speech act with associated semantics and protocol (see also Finin, Labrou & Mayfield, 1997).

Ontologies are defined as controlled vocabularies providing the basic terminology necessary for representation of a domain of discourse (Gruber,

1999). Ontologies are considered the basic elements of knowledge media. A knowledge medium is defined as "*a computational environment in which explicitly represented knowledge serves as a communication medium among people and their program*" (Gruber, Tenenbaum & Weber, 1992).

Given a common communication language, heterogeneous agents can communicate with each other and build federations of agents that strive to achieve a common goal, which requires the combined knowledge of all participating agents (Figure 1).

The proposed ACL was successfully applied in several projects (see Finin et al., 1994), and the language is at present a de-facto standard in the artificial intelligence scientific community. The proposed concepts for an ACL have influenced in great manner the standard for an ACL proposed by the Foundation for Intelligent Physical Agents (FIPA) see http://www.drogo.cselt.stet.it/fipa) and have inspired similar projects dedicated to the development of specific ontologies (see for example Fensel et al., 1999). At present further developments of the Stanford ACL are rivaled by increasingly sophisticated Internet and Knowledge Markup Languages based on the Extended Markup Language (XML) (Bray, Paoli, Sperberg-McQueen, & Maler, 2000), which can also be used by agents.

In summary, the Stanford approach for knowledge media has concentrated on solving the first two tasks necessary to build knowledge media mentioned by Stefik and described in the section "The Origin of the Knowledge Media Concept," i.e., on the development of a specific standardized communication language for distribution and exchange of knowledge between artificial agents.

Figure 1: A federation of agents communicating through ACL (Genesereth & Ketchpel, 1994)

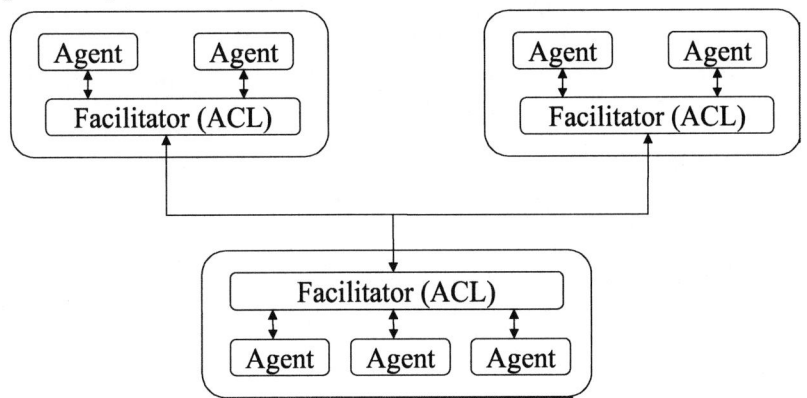

The Concept of Knowledge Media as Proposed by Schmid

The concept of knowledge media proposed by Schmid (1997a) is grounded in Stefik's (1986) vision, but extends the Stefik and Stanford approach for building knowledge media in three directions: First, it not only considers expert systems, but generalizes knowledge sharing for all kinds of knowledge sources. Second, it explicitly includes the agents who create and use the knowledge medium by introducing the concept of community. Thus it extends the concept of knowledge media towards knowledge ecologies– systems consisting of human and artificial agents sharing and creating knowledge within digital media (see also Davenport, 1997; Ginsburg, 2000). Third, it explicitly considers not only agent communication processes but any kind of organizational structure and coordination processes required and applied by the community creating and exchanging knowledge.

Definition and Components of Knowledge Media

Schmid (1997) defines a knowledge medium as a platform providing a sphere for the management and exchange of knowledge within a confined community of human and artificial agents. These knowledge media provide metaphors of meeting places providing communication channels and common knowledge repositories.

According to Schmid, a knowledge medium is comprised of the following components (for a detailed and formalized derivation see Lechner and Schmid, 1999; Schmid, 1997a, 1997b; Schmid and Stanoevska, 1998:

- a *logical space*, which defines the common language applied by the medium as well as the common syntax and semantics of the knowledge represented in and managed by the medium.
- a *system of channels*, i.e., (interactive) carriers capable of storing and transporting knowledge over space and time.
- an *organizational structure* describing the coordination and cooperation models applied by the medium in terms of roles, behavioral rules for agents as well as processes implemented for knowledge creation, sharing and usage.

In the following section, the components of a knowledge medium will be explained in more detail in accordance with Stanoevska-Slabeva and Schmid (2000).

Logical Space

Language is the necessary prerequisite for the externalization and exchange of knowledge and a binding element in a community. The logical

system of a medium denotes the language, which is used for capturing and communication of knowledge. It may have different forms from natural language to a formal
representation of the domain of discourse. We can take advantage of the ability of the
digital medium to enhance human intelligence and reasoning only by providing a formal
representation of the domain of discourse. The formal representation might be an expert
system or it might, for example, provide a meta-layer over the content and channels of
the medium in the form of an ontology (Gruber, 1999), enabling classification and
structuring of content as well as allowing for its automatic retrieval and combination.

System of Channels

Channels are a system of information carriers connected to a network of
transportation means for information. Channels are the carriers containing
explicit knowledge, which is exchanged between the communicating agents.
A *carrier* includes a physical medium (sound waves in the case of human
speech) and applies a language (syntax and semantics) for the logical
representation of the contents.

Carriers can provide some interactivity. *Interactivity* means that the
carrier can respond to some extent to inputs and inquiries. This feature
distinguishes it from traditional channels like paper documents.

Interconnected channels provide a communication space, which can be
used by agents to exchange knowledge. Filled with content, channels turn to
interactive, ubiquitous, and multimedia information objects. The most preva-
lent information objects in companies are electronic documents.

Organizational Structure of the Medium

As mentioned above a medium is a metaphor of a meeting place
providing communication channels and common knowledge repositories.
The entrance into the space, i.e., the usage of the medium, is usually subject
to certain rules defined by the community. Such rules can be implemented on
the medium as part of its organizational structure. Further components of the
organizational structure of a medium are: roles, protocols, and processes.
Roles define the required capabilities, rights, and obligations of agents
participating in the community and using the knowledge medium. *Protocols*
are a set of rules that govern the interaction of agents. *Processes* describe the
sequence of tasks required for knowledge creation, management and sharing.

The Relationship Between the Components of the Knowledge Medium

The above described components of media are interwoven with each
other and form an entity of human and artificial agents connected through a
semantic space facilitating communication (Figure 2). Communities of

Figure 2: The components of knowledge media (Schmid, 1997a)

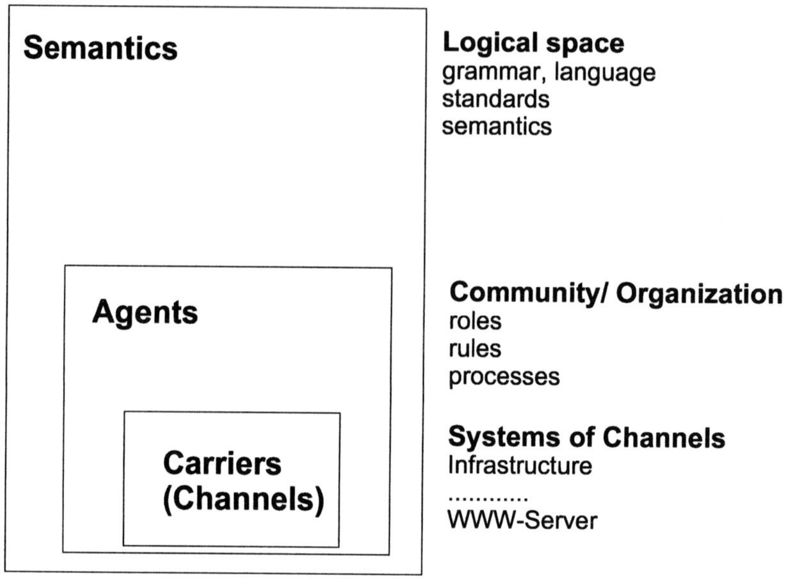

Semantics	**Logical space** grammar, language standards semantics
Agents	**Community/ Organization** roles rules processes
Carriers (Channels)	**Systems of Channels** Infrastructure .... WWW-Server

agents, striving towards a common aim, employ the above described organized channel structure to create, store, manage and share knowledge.

Different communities can model any given world using different languages. If a semantic connection between the resulting knowledge media, i.e., between their logical spaces, can be established, the exchange of explicit knowledge between them can be facilitated by automated translation from one language into another. In this case, the different knowledge media form a knowledge media *net*. An example of a knowledge media net for the scientific community is described by Handschuh et al. (1998).

Other Approaches for Building Knowledge Media

Besides the above mentioned initiatives for developing concepts for knowledge media, there are others that emphasize different perspectives of knowledge media. An interesting perspective is the one by the Knowledge Media Design Institute at the University of Toronto in Canada. Here knowledge media are defined as "the building blocks of a learning society" and are characterized by the following features (Baecker, 1997):

- Knowledge media are computational artifacts, which incorporate both data and processes.
- They can compute new facts and can configure and present information in a unique way based in part on roles built into the medium.
- They incorporate both task spaces in which people carry out their work and

interpersonal spaces, i.e., environments or frames, in which communication takes place.

- The central purpose of knowledge media is to help communities of individuals to think, communicate, learn, and create knowledge.

This approach reflects the basic features of knowledge media proposed by Stefik (1986; for example, computational artifacts comprising data and processes and configuration and presentation of information) even though it is not directly grounded in its vision. In addition, it considers communities as users of a knowledge medium and differentiates between personal and interpersonal spaces, pointing to the need of a personalized usage of a knowledge medium.

Another initiative is the one by the Knowledge Media Institute at the Open University in London, Great Britain, http://kmi.open.ac.uk/home-f.html, where knowledge media are defined as "the capturing, storing, imparting, sharing, accessing, and creating of knowledge" (Eisenstadt, 1995). The Knowledge Media Institute develops knowledge media for special communities as well as technologies for knowledge media. Knowledge media are seen as platforms for lifelong learning. Thereby, meta-data based on XML (see, for example, Buckingham-Shum, 1998) as well as ontologies (see Domingue, 1998; Motta, Buckingham-Shum, and Domingue, 1999) are considered an important approach for semantic enrichment of the Internet and a necessary prerequisite for an efficient use of the available knowledge sources on the Internet.

INTERNET–THE FUTURE GLOBAL KNOWLEDGE MEDIUM?

Since its appearance, the Internet has evolved to the world's biggest digital library. As a result of its free publishing policy, it is comprised of various information sources, such as teaching materials, on-line dictionaries and encyclopedias, scientific publications, newspapers, magazines, company reports, stock reports, and much more. As a result, the Internet has evolved into a valuable source of knowledge in the everyday life of many people. It is often called the global knowledge medium or the next knowledge medium.

However, if we take Stefik's (1986) vision for the next knowledge medium, does the Internet really meet all the requirements of a knowledge medium? In order to answer this question the features of the Internet will be compared to the defining features of knowledge media as proposed by Stefik (1986) and described in the section "The Origin of the Knowledge Media Concept."

The Internet certainly is a global body of knowledge available on the numerous servers. However, currently they are isolated nodes of knowledge and an integrated search and combination of knowledge is limited. The second basic component of knowledge media–a common or standard communication language–is also available. However, the third component–a common vocabulary or common vocabularies of specific communities–are not available yet. As a result, there is a language barrier impeding the global use of available knowledge resources. This is also the main obstacle preventing the Internet from becoming a true global knowledge medium. As long as there is no common terminology, an integrated search and automatic combination of knowledge as well as the definition of generic processes for knowledge creation, management, and sharing will not be possible.

Another obstacle that stands in the way of the Internet becoming the global knowledge medium is the lack of machine-readable representation of the available resources on the Internet (Berners-Lee, 1998; Stefik, 1999). Therefore, intelligent inference mechanisms can not be applied to facilitate knowledge combination and reuse. Consequently, the Internet can evolve into the next knowledge medium, i.e., it can provide an "intelligent" and global knowledge infrastructure, only if common languages are developed, which will enable interoperability between available knowledge sources.

The initiative named "Semantic Web" and led by Tim Berners-Lee (http://www.w3.org/DesignIssues/Semantic.html) provides a first step towards achieving this goal. The aim of the initiative is the development of at least syntactic and to a certain extent semantic agreements for the annotation of available information on Internet based on the XML and RDF (Resource Description Framework) technology (for detailed information on RDF see http://www.w3.org/RDF/). The application of RDF will hopefully result in common terminology at least for specific communities, so that the knowledge sources available on the Internet can be combined in the sense of a knowledge medium.

SUMMARY AND OUTLOOK

In 1986 Stefik introduced the concept of knowledge media based on the interactive and "intelligent" digital medium. He envisioned that the integration and interoperation of specialized expert systems would provide a new intelligent communication and transportation infrastructure, which will enable fast diffusion of knowledge. Basic elements of knowledge media are standardized communication and knowledge representation languages.

Several initiatives have taken over Stefik's vision of knowledge media and have developed special solutions for their implementation. The solutions show that the basic idea referring to expert systems as building blocks for knowledge media can be extended to any digital knowledge source. As a result, we can observe the clear development of the concept of knowledge media in the following directions:

1. A changing role of the medium from a self-reasoning network of expert systems to an interactive communication and knowledge exchange space comprising different types of external knowledge, for example, documents.
2. Development of common languages within defined communities and translations between languages of communities. It will probably be impossible to develop a common language for any domain and for all communities.
3. One important part of knowledge is the tacit knowledge of humans. Thus, knowledge media should be considered as integrated systems of artificial and human knowledge. This means a clear shift from self-reasoning interoperable expert systems towards knowledge ecologies providing a symbiosis between artificial and human agents and knowledge sources. Communities are the organizational forms for such knowledge ecologies.

The third feature of Stefik's (1986) vision–the development of a knowledge market–has not been addressed by any of the approaches included in this paper. This is due to the fact that first standardized communication languages and knowledge representation languages are needed, which will render knowledge comparable and will provide the foundation for implementation of market coordination mechanisms.

ACKNOWLEDGMENTS

The author would like to thank Brigette Buchet, executive editor of *The International Journal of Electronic Commerce and Business Media* for her help in maintaining a high quality of English throughout.

REFERENCES

American Heritage Dictionary of the English. (2000). On the World Wide Web at: http://www.dictionary.com/cgi-bin/dict.pl?term=Media.

Armour, P. G. (2000). The case for a new business model–Is software a product or a medium. *Communication of the ACM*, 43(8), 19-22.

Baecker, R. (1997). *The Web of Knowledge Media Design*. Retrieved December 20, 2000 on the World Wide Web: http://www.kmdi.org/whatskmdi.htm.

Berners-Lee, T. (1998). *Semantic Web Roadmap*. Retrieved on the World Wide Web: http://www.w3.org/DesignIssues/Semantic.html.

Bray, T., Paoli, J., Sperberg-McQueen, C. M. and Maler, E. (2000). *Extensible Markup Language (XML)* 1.0 (2nd ed.). W3C Recommendation, October. Retrieved February 15, 2001 from the World Wide Web: http://www.w3.org/TR/REC-xml.

Buckingham-Shum, S. (1998). Evolving the Web for scientific knowledge: First steps towards an "HCI knowledge Web". *Interfaces, British HCI Group Magazine*, 39, 16-21.

Davenport, T. (1997). *Information Ecology–Mastering the Information and Knowledge Environment*. New York: Oxford University Press.

Domingue, J. (1998). *Tadzebao and WebOnto: Discussing, Browsing and Editing Ontologies on the Web*. (Report No. KMI-TR-69). Open University: Knowledge Media Institute.

Drucker, P. F. (1991). *The New Reality: Econ*.

Eisenstadt, M. (1995). *The Knowledge Media Generation*. Retrieved February, 2000, on the World Wide Web: http://kmi.open.ac.uk/kmi-misc/&kmi-feature.htm.

Fensel, D., Angele, J., Decker, S., Erdmann, M., Schnurr, H. P., Staab, S., Studer, R. and Witt, A. (1999). On2broker: Semantic-based access to information sources at the World Wide Web. *Proceedings of the World Conference on the WWW and Internet (WebNet 99)*, Honolulu, Hawai, US, 25-30.

Finin, T., Fritzson, R., McKay, D. and McEntire, R. (1994). KQML as an agent communication language. *Proceedings of the Third International Conference on Information and Knowledge Management (CIKM'94)*, ACM Press.

Finin, T., Labrou, Y. and Mayfield, J. (1997). KQML as an agent communication language. In Bradshaw, J. (Ed.), *Software Agents*. Cambridge: MIT Press.

Genesereth, M. R. and Ketchpel, S. P (1994). Software agents. *Communications of the ACM*, 37(7), 48-53.

Ginsburg, M. (2000). Internet document management systems as knowledge ecologies. In Sprague, E. (Ed.), *Hawaiian International Conference on System Sciences (HICSS 2000)*, IEEE Press.

Gruber, T. (1993). Toward principles for the design of ontologies used for knowledge sharing. *Technical Report KSL 93-04. Knowledge Systems Laboratory*, Stanford University. Retrieved September, 1999, on the World Wide Web: http://ksl-Web.stanford.edu/KSL_Abstracts/KSL-93-04.html.

Gruber, T. (1999). *What is an Ontology?* Retrieved April 22, 1999 on the World Wide Web: http://www-ksl.stanford.edu/kst/what-is-an-ontology.html.

Gruber, T., Tenenbaum, J. M. and Weber, J. C. (1992). Towards a knowledge medium for collaborative knowledge development. In Gero, J. S. (Ed.),

Proceedings of the Second International Conference on Artificial Intelligence in Design, Pittsburg, USA, June 22-25, 413-431. Boston: Kluwer Academic Publishers.

Handschuh, S., Lechner, U., Lincke, D. M., Schmid, B. F., Schubert, P., Selz, D. and Stanoevska, K. (1998). The NetAcademy-A new concept for on-line publishing and knowledge management. In Margaria, T., Steffen, B., Rückert, R. and Posegga, J. (Eds.) *Services and Visualization–Towards a User-Friendly Design*, 29-43. Berlin/ Heidelberg: Springer Verlag.

Hill, A. and Watson, J. (1997). *A Dictionary of Communication and Media Studies*. London: Arnold.

Labrou, Y. and Finin, T. (1997). A proposal for a new KQML specification. Baltimore: *Technical Report No. TR CS-97-03 of the Computer Science and Electrical Engineering Department*, University of Maryland, Baltimore. Retrieved September, 1999, on the World Wide Web: http:// www.cs.umbc.edu/kqml/papers/.

Lechner, U. and Schmid, B.F. (1999). Logic for media–The computational media metaphor. In Sprague, R. H. (Ed.), *Proceedings of the 32nd Hawaii International Conference on System Sciences (HICSS-32)*. Los Alamitos, CA: IEEE Computer Society.

McLuhan, M. (1964). *Understanding Media*. NY: McGraw-Hill.

Motta, E., Buckingham-Shum, S. and Domingue, J. (1999). Case studies in ontology-driven document enrichment. In *12th Workshop on Knowledge Acquisition, Modeling and Management (KAW '99)*. Banff, Alberta, Canada.

Nonaka, I. (1991). The knowledge creating company, *Harvard Business Review*, November-December, 96-104.

Nonaka, I. and Konno, N. (1999). The concept of "Ba": Building a foundation for knowledge creation. *California Management Review*, 40(3), 40-54.

On-line Encyclopedia dedicated to Computer Technology. (2000). Retrieved on the World Wide Web: http://Webopedia.Internet.com/TERM/m/ media.html.

Probst, G., Raub, S. and Rohmhard, K. (1997). *Wissensmanagement–Wie Unternehmen ihre Wertvollste Ressource Optimal Nutzen*. Wiesbaden: Gabler Verlag.

Schmid, B. F. (1997a) The concept of media. In *Workshop on Electronic Markets Workshop*. Maastricht, The Netherlands.

Schmid, B. F. (1997b). Wissensmedien: Konzept und schritte zu ihrer realisierung. *Working Paper, Draft No. 01*, Institute for Media and Communications Management, University of St. Gallen, Switzerland, December 1997. Retrieved February 15, 2001 on the World Wide Web: http://www.netacademy.org/netacademy/publications.nsf/all_pk/1212.

Schmid, B. F. and Stanoevska, K. (1998). Knowledge media: An innovative concept and technology for knowledge management in the information age. *Proceedings of the 12th Biennal International Telecommunications Society Conference - Beyond Convergence*. Stockholm, Sweden: IST'98.

Schmid, B. F., Eppler, M. J., Lechner, U., Schmid-Isler, S. B., Stanoevska-Slabeva, K., Will, M. and Zimmermann, H. D. (1999). A glossary for the NetAcademy: Issue 1999. *Working Report 1999-08 mcm institute.* Retrieved on the World Wide Web: http://www.businessmedia.org/netacademy/publications.nsf/all_pk/1296.

Stanoevska-Slabeva, K. and Schmid, B. (2000). Community supporting platforms. *Proceedings of the 33rd HICSS Conference*, Hawaii.

Stefik, M. J. (1986). The next knowledge medium. *AI Magazine*, 7(1), 34-46.

Stefik, M. J. (1999). *The Internet Edge.* Cambridge, MA: The MIT Press.

Wagner, D.N. (2000). Software agents take the Internet as a shortcut to enter society: A survey of new actors to study for social theory. *First Monday.* Retrieved on the World Wide Web: http://www.firstmonday.dk/issue/issue5_7/wagner.html.

Chapter II

Experience With a Functional-Logic Multi-Agent Architecture for Medical Problem Solving

Giordano Lanzola
University of Pavia, Italy

Harold Boley
German Research Center for Artificial Intelligence, Germany

INTRODUCTION

There exist various experiences in constructing applications in functional and logic languages alone as reported, e.g., in the conference series *Lisp and Functional Programming* and *Practical Applications of Prolog*, respectively. Applications in integrated functional-logic languages (Hanus, 1994) have been less frequent, however, and little has been reported about experiences with them. Historically, functional-logic languages started off with practically oriented, loose couplings of LISP and PROLOG components such as LOGLISP, and LISP-based hybrid expert-system shells often contained some PROLOG-like component. In the few applications written in these LISP-PROLOG couplings and shells, however, one of the integration partners was normally dominating. Later, theoretically oriented, tight syntheses of functional and logic components such as EQLOG were studied. Even in those syntheses that were really implemented, very few applications were developed–apart from the published examples demonstrating the new paradigm. Later, the accumulated theoretical insights in functional-logic concepts and implementations have led to a new treatment of practical problems, as attempted, e.g., by ALF, BABEL, and RELFUN. For the problems attacked, an initial distinction between internal tools (e.g., the ALF compiler in ALF) and external applications (e.g., the RTPLAST selector in RELFUN) can be made. Application-oriented projects in whose context RELFUN has been developed have exerted some pressure to test its general concepts on the latter category, namely on real-world examples from technical modeling, mostly in mechanical engineering. Experiences with RTPLAST, a RELFUN knowledge base for selecting Recyclable ThermoPLASTtics that satisfy engineering requirements, have for example been reported at a materials-science conference (Boley et al., 1994).

The present paper reports on our experience with applying RELFUN (Boley, 1999) to problems in distributed medical care. This application arose "externally" in the original sense of the word: Massimiliano Campagnoli, working with the LISP-based expert-system shell KEE, supporting frames as well as forward and backward rules, noticed RELFUN on the net and switched to it, since frames are mappable into clauses and RELFUN's rules offer more versatility than KEE's. After his initial implementation of a RELFUN-based distributed medical-care system, he contacted the second author, further developing RELFUN, and the Pavia/Kaiserslautern teams joined forces, with the first author also being the expert in the medical domain.

Massimiliano Campagnoli and the authors then set out together on the implementation of a system composed of multiple communicating agents aimed at exploiting the rapidly emerging methodologies addressing computer supported cooperative work (CSCW) for the management of patients affected by acute myeloid leukemia (AML). AML was carefully selected as the application domain since it greatly emphasizes the need for different healthcare professionals to cooperate towards the achievement of a common goal. We first extended RELFUN with a powerful set of KQML (Finin et al., 1992) communication primitives, and then we implemented several agents, each one encapsulating the domain specific and the strategic knowledge belonging to a different professional. The overall prototype described in this paper illustrates the potential of a methodology which may also prove to be useful in improving the quality and reducing the cost-efficiency ratio of the healthcare delivery process by laying down new organizational infrastructures exploiting distributed information technology.

The following aspects turned out to be important in this application (reasons known in advance, hence critical in choosing this language, are emphasized by an "*"):

* Like the previously used systems KEE and KQML, RELFUN is written in LISP, so that its implementation was easy to interface with existing software;
* RELFUN is highly portable among COMMON LISP versions and easy to install, especially if only the interpreter is needed, as in this prototype;
• Kernel RELFUN is a simple, if not minimal, functional-logic integration, hence it is also easy for non-computer scientists to learn also;
* The on-line documentation is satisfactory, containing many executable examples on which other programs can be modeled;
• The RELFUN implementation consists of an interpreter, for development, as well as several source-to-source translators and a WAM compiler, for delivery (with two emulators, one written in LISP, the other in C (Perling, 1998);
• We have practiced a flexible user-supporter relationship as part of which several new RELFUN features were developed (e.g., dynamic assertz and retractx).

These existing kernel RELFUN features were needed:

* Relations and functions, both essential in the application, are integrated in a natural manner;
* Relations and functions may be of varying arity, helping to define generic KQML ask messages;
* Many functions can be inverted, as convenient for sharing the specification between two (counter-)directed computations;
* Higher-order functions permit a declarative implementation of messages passed as data and activated in their receivers (function variables) as well as a high-level encoding of agent abilities within facilitators (functional values);
* Finite domains are usable as first-class citizens, allowing a compact and efficient representation of disjunctive values.

APPLICATION-ORIENTED FUNCTIONAL-LOGIC CONCEPTS

The two classical declarative paradigms are functional programming in the tradition of pure LISP and logic (relational) programming following pure PROLOG. Basic communalities between these or modern functional (e.g., Haskell) and relational (e.g., Godel) languages have led to increasing efforts at their integration in search of a unified declarative paradigm for "functional-logic" or "relational-functional" programming (cf. the survey; Hanus, 1994). Here we discuss application-oriented issues of such an integration with an emphasis on syntactic and semantic simplicity as called for by practical use. Since the medical application is based on RELFUN (Boley, 1999), this language will be used to illustrate the required functional-logic concepts.

Having the new infix "&" precede an explicit value to be returned, as a simple extension to PROLOG's Horn-clause syntax, a RELFUN operator is defined by a system of *valued clauses* which may embody increasing expressive power as follows:

(1) *Unit clauses* (facts) op(arg_1,...) for extensional relation definitions are implicitly `true`-valued. For example, healthy(p0) is a ground fact stating the health of individual p0. Generalizing PROLOG's "|"-use in lists, varying arity is permitted in clause heads and in calls, e.g., the calls healthy(Who), healthy(First|Rest), and healthy(|Args) respectively bind Who to p0, First and Rest to p0 and [], and Args to [p0].

(2) *Unconditional equations* op(arg_1,...) :& exp for operator (normally, function) definitions whose case distinctions are made only via unification of left-hand sides (written as clause heads) return an explicit right-hand-side (rhs) expression value; in *molecular rules* exp specializes to a term. For example, bodytemp(p1) :& 39 is a ground molecular rule, specializing exp to a numeric constant, which specifies a unary bodytemp function that maps individual p1 to value 39. The calls bodytemp(p1) and bodytemp(Who) now both return 39, the latter also binding Who to p1.

(3) *Non-unit clauses* (rules) op(arg_1,...) :- cnd_1,...for relation definitions whose case distinctions require conditions (written as clause premises, also used for the accumulation of partial results) are implicitly `true`-valued. For example, bodytemp(P,37.5) :- healthy(P) specifies a binary bodytemp relation without preferred direction of computation. Using the fact in (1), the call bodytemp(p0,V) binds V to 37.5.

(4) *Conditional equations* op(arg_1,...) :- cnd_1,...& exp for operator (normally, function) definitions whose case distinctions require conditions, as in (3), again return an explicit rhs expression value, as in (2). For example, bodytemp(P) :- healthy(P) & 37.5 specifies a unary bodytemp function that maps variable P to constant 37.5. Using the fact in (1), the call bodytemp(p0) returns 37.5. Alternatively, bodytemp(P,37.5) :- healthy(P) & 0.9 specifies a binary bodytemp function that maps variable P and constant 37.5 to probability 0.9. Now the call bodytemp(p0,V) binds V to 37.5 and returns 0.9.

RELFUN's primary concern is *minimal integrative extensions* of both declarative programming paradigms wrt fusion into a relational-functional kernel (Boley, 1999). For this, two relational essentials, (R1) and (R2), are mapped to corresponding extensions of functional programming, and two functional essentials, (F1) and (F2), are similarly transferred to the relational paradigm:

(R1) The relational essential of permitting first-order *non-ground terms* (terms being or containing free logic variables, which may become bound by calls) is transferred to functional programming: A function can take non-ground terms as arguments by using (two-sided) unification instead of (one-sided) matching, and similarly can return non-ground terms as values. With call-by-value evaluation of functional applications this will lead to innermost conditional narrowing. Many functions (such as the unary bodytemp function) can thus be called inversely using the ".=" primitive, which generalizes PROLOG's "is" to user-defined functions.

(R2) Since a non-ground (e.g., inverse) function call may deliver several "solution values," this also entails a transfer of the relational essential of *(don't-know) non-determinism* (solution search, implemented by backtracking enumeration as in PROLOG) to functional programming. Historically, however, non-ground functional programming was proposed as a result of relational-functional integrations, while non-deterministic functional programming was first introduced as a purely functional generalization. For finitely non-deterministic calls RELFUN's ordered `bagof` version `tupof` can be used to *return* the list of all values. For example, with the sequence of clauses in (1), (2), and (4) `tupof(bodytemp(Who))` returns `[39,37.5]`.

(F1) The functional essential of *application values* (function applications return value terms, hence can be nested into "functional compositions") is transferred to relational programming: A relation whose call succeeds always returns the value `true` in the manner of a characteristic function, besides possibly binding variables. On the other hand, each argument of a relation call may be the value returned by an application rather than a directly specified term. Hence, as a constructor-based language, RELFUN explicitly distinguishes (passive, instantiated) structures from (active, evaluated) applications, marking instantiation vs. evaluation by square brackets vs. round parentheses. For example, with (1) and RELFUN's prelude definition `tup( |R) :& [ |R]`, the nesting `healthy(first(tup(p0,p1)))` returns `true`.

(F2) The functional essential of *higher-order functions over named functions* (named functions as functional arguments and values) is transferred to relational programming: A relation can take named relations as arguments. Our relational sublanguage permits logic variables also in the relation position of queries and definitions ("predicate variables"), introducing some syntax of *second-order predicate logic*; this is transferred to functional programming ("function variables"). The higher-order operators permitted here, although practically very useful, embody just conservative, syntactic extensions that can be reduced to first-order versions via an `apply` dummy operator. For example, with (1), the call `Pred(p0)` binds `Pred` to `healthy`, and, using this binding, `Pred(p1)` fails.

On the basis of the integrated RELFUN kernel provided by (R1) to (F2), common functional-logic extensions were developed. To illustrate, *first-class finite domains* are introduced as `dom` terms, which can be bound to variables, used as arguments, embedded in data structures, and returned as values. For successful constant-domain unification the constant must occur among the domain's constants; on domain-domain unification the intersection of the domains is computed. For example, both `p1 .= dom[p0,p1,p2]` and `dom[p0,p1] .= dom[p1,p2]` return `p1`, equivalent to `dom[p1]`.

A MULTI-AGENT ENVIRONMENT IMPLEMENTED ON TOP OF RELFUN

Solving real problems is a complex cognitive activity requiring the exploitation of different knowledge types and calling for the dynamic selection of the goal to pursue as well as the subsequent choice of the most appropriate reasoning technique given that goal and the context shown by the problem. Thus a great deal of research in AI has been devoted to the identification of suitable knowledge representation formalisms, reasoning techniques, as well as to formulating epistemological frameworks (Ramoni et al., 1992) and computational architectures supporting them (Lanzola & Stefanelli, 1993).

Nevertheless, the key issue is that applications featuring a high complexity level necessarily require the encapsulation of increasing amounts of knowledge up to the point that the costs concerned with the process of acquiring, representing and maintaining that knowledge actually become unbearable also in research contexts (Lenat & Feigenbaum, 1991). To overcome those limitations, the scientific community has become increasingly aware of the need of developing sharable and reusable bodies of knowledge which may be easily interchanged among different systems (Klinker et al., 1991; Neches et al., 1991). However many authors are actually considering that the ultimate solution towards automating problem solving entails a full reconsideration of its paradigm. Therefore, instead of adopting a "monolithic" approach where all of the problem solving expertise is constrained into a single system, the current efforts envision a different architecture composed of a community of multiple cooperating agents, each one responsible of providing a different expertise (Fisher & Wooldridge, 1993; Jennings et al., 1993).

This approach is much closer to the way problems are actually solved in real life through the cooperation of multiple individuals. Furthermore, it is also helpful in overcoming the constraint of integrating, within one single implementation, different knowledge types which presumably have been independently developed and very often make use of different, if not incompatible, formalisms. Several research efforts are currently aimed either at developing generic frameworks for building agents like AOP (Shoham, 1993) or analyzing the pragmatic issues concerned with inter-agent transactions as KQML (Finin et al., 1992). While the former ones are more oriented towards representing agents by modeling their internal states in terms of beliefs, desires and intentions, the latter ones limit themselves to the investigation of the communication primitives as expressed by speech act theory (Searle, 1969). In view of experimenting with a multi-agent architecture applied in clinical medicine, we preferred the KQML approach since it allowed us to focus on the communication requirements needed for exchanging and interpreting messages among separate agents. This has been useful in complementing the knowledge representation capabilities typical of a functional-logic programming approach such as the one provided by RELFUN, which was being used for implementing our agents.

KQML (Knowledge Query and Manipulation Language) is a language and a protocol designed to support communication between separate knowledge-based modules as well as between knowledge-based systems and databases. It has been developed within the ARPA Knowledge Sharing Effort (KSE; Neches et al., 1991), a consortium established to develop conventions facilitating the sharing and reuse of knowledge. Based on speech act theory, it provides for a message format and a message-handling protocol supporting runtime interaction among different agents, while using standard messaging protocols as transport layer, e.g., SMTP, HTTP, etc. A typical KQML message consists of a performative with its associated content, expressed in a suitable language, and a set of transport arguments. Those basically include a message sender, a receiver, and two labels, one for properly identifying the message (identifying label) and the other one to be used as the message identifier in a possible peer agent reply (expected label). Other arguments indicating the assumed ontology and the content language can be added as well.

Several reserved performatives are defined in KQML:
- basic informative performatives (tell, untell and deny);
- database performatives (insert, delete, delete-one, delete-all);
- basic query performatives (evaluate, reply, ask-if, ask-about, ask-one, ask-all, sorry);

- multi-response query performatives (stream-about, stream-all, eos);
- basic effector performatives (achieve, unachieve);
- generator performatives (standby, ready, next, rest, discard, generator);
- capability-definition and notification performatives (advertise, subscribe, monitor);
- networking performatives (register, unregister, forward, broadcast, pipe, break, transport-address);
- facilitation performatives (broker-one, broker-all, recommend-one, recommend-all, recruit-one, recruit-all).

The performative's content can be expressed in different languages: KIF, PROLOG, LISP and RELFUN, to mention some. Of course, the agent implementation language and the performative's content language do not need to be the same; nevertheless, using RELFUN both to implement the agent and to encode the performative's content is helpful in easing and speeding up the processing of incoming messages and providing appropriate replies.

Many of the KQML performatives described by Finin et al. (1992) have been therefore implemented using RELFUN's capability to process declarative knowledge. Here we give just a few simplified examples, skipping over network details.

The general form of a KQML message encoded in RELFUN is as follows:

```
kqml-msg(Sender,Receiver,Performative,Content,
   In-reply-to,Reply-with).
```

where `Sender` and `Receiver` are the message sender and receiver, respectively. `Performative` is one of the reserved KQML performative names as defined by Finin et al. (1992). `Content` is a normally non-ground list of the form `[Result,Finding[Arg1,...,ArgN]]` where `Result` is obtained by activating the `Finding` structure to a function call via `Result .= Finding(Arg1,...,ArgN)`. `Finding` is any kind of medical findings such as body temperature, blood pressure, etc., `In-reply-to` is the identifying label in a reply, while `Reply-with` indicates whether the sender expects a reply, and if so its expected label.

To send a KQML message, a call to:

```
send-msg (Sender,Receiver,Performative,Content,
   In-reply-to,Reply-with).
```

provides for the actual message delivery from `Sender` to `Receiver` using TCP/IP as the transport protocol. `Receiver` monitors its input message queue until a message arrives and then asserts the message in its knowledge base.

For example, a message sent from a medical agent `a1` (say, an internist) to a medical agent `a2` (say, a nurse) representing a query about the `bodytemp` value for patient p1 might be expressed with the following fact, asserted in a2's knowledge base:

```
kqml-msg(a1,a2,ask-one,[What,bodytemp[p1]],_,r1).
```

where a1 is the sender, a2 the receiver, ask-one the KQML performative, [What,bodytemp[p1]] the content, and r1 a label for the expected reply. Agent a1 keeps tracks of any sent message asserting them in its knowledge base. Agent a2 processes this message using the rules:

```
process-msg() :-
    kqml-msg(Sender,a2,ask-one,[Result,Finding[ |Args]],_,
    Reply-with),
    Result .= Finding( |Args),
    send-msg(a2,Sender,tell,[Result,Finding[ |Args]],
    Reply-with,_).
process-msg() :-
    kqml-msg(Sender,a2,ask-one,_,_,Reply-with),
    send-msg(a2,Sender,sorry,_,Reply-with,_).
```

The first process-msg rule leads to the reply:

```
kqml-msg(a2,a1,tell[39,bodytemp[p1]],r1,_).
```

being asserted in a1's knowledge base, provided that the molecular rule:

```
bodytemp(p1) :& 39.
```

is present in a2's functional bodytemp definition. Otherwise the second process-msg rule sends a sorry performative to agent a1, which indicates that a2 understands but is not able to provide any response to the message.

If agent a1 wants to know the name of some patient showing a given body temperature value, say 39, it will send a message to a2, who, once having received a1's query, will assert the fact:

```
kqml-msg(a1,a2,ask-one,[39,bodytemp[Who]],_,r1).
```

and, afterwards, will process it using the same process-msg rule employed above, with 39 .= bodytemp(Who) instantiating Who to p1. RELFUN's invertible function bodytemp and higher-order constructor Finding thus help in answering different queries using the same specification.

The names of all patients showing a body temperature value equal to 39 can be asked to a2 as:

```
kqml-msg(a1,a2,ask-all,[39,bodytemp[Who]],_,r1).
```

and answered by the rule:

```
process-msg() :-
    kqml-msg(Sender,a2,ask-all,[Result,Finding[ |Args]],_,
    Reply-with),
    Result .= tupof(Finding( |Args)),
    send-msg(a2,Sender,tell,[Result,Finding[ |Args]],
    Reply-with,_).
```

Database performatives like `insert` and `delete` are easily implemented via
   RELFUN's `assertz` and `retractx`:

```
process-msg() :-
    kqml-msg(Sender,a2,insert,[Result,Finding[ |Args]],_,
      Reply-with),
    assertz(Finding( |Args) :& Result.).
process-msg() :-
    kqml-msg(Sender,a2,delete,[Result,Finding[ |Args]],_,
      Reply-with),
    retractx(Finding( |Args) :& Result.).
```

A KQML facilitator is a specialized agent knowing what other agents are able
to do. For example, our medical facilitator knows that a2 is able to answer queries
on body temperature values, as well as on many other findings, keeping a table like:

```
ability(a2)  :& bodytemp.
ability(a2)  :& weight.
ability(a2)  :& age.
. . .
```

Note that the values `bodytemp`, `weight`, `age`, ... of the ability higher-
order function are themselves function names, representing a2's "functionality."

If an agent, say a3, wants to acquire the body temperature value for
patient p1, but does not know who can provide this information, it can contact
the facilitator via the message:

```
kqml-msg(a3,facilitator,broker-one,
    [bodytemp,Agent],_,r1).
```

The facilitator, in turn, will determine agent a2 as the appropriate
provider via the inversion `bodytemp .= ability(Agent)`:

```
process-msg() :-
    kqml-msg(Sender,facilitator,broker-one,
     [Ability,Agent],_,Reply-with),
    Ability .= ability(Agent),
    send-msg(facilitator,Sender,tell,
     [Ability,Agent],Reply-with,_).
```

The determination of the values of the body temperature and ability
functions was possible by a simple table lookup. In general, working out
a value may require much more complex tasks, where further agents
implementing additional reasoning and planning may be involved. This
occurs, for instance, when a query is aimed at finding out by which disease
a patient is affected.

THE ACUTE MYELOID LEUKEMIA
TASK PLANNING

Medicine has always played a key role among the fields chosen by AI researchers for implementing prototypical applications aimed at testing any new methodology being developed. This is due to several distinguishing aspects imposing different constraints and increased demands on medical applications, thereby setting them aside from those addressing any other domain (Blois, 1984). In particular there is hardly a domain featuring a more widespread distribution of knowledge, expertise and data as medicine does. Managing patients is a complex cognitive activity requiring high interoperability among the several healthcare professionals involved, who contribute with diverse knowledge support and must be able to exchange information and share a common understanding of a patient's clinical evolution. Different users will then require different support types from different applications, which will be exploited for helping them in accomplishing their daily tasks (Lanzola et al., 1996).

Part of our research work is therefore being aimed at demonstrating the potential of combining the expressive power of a functional-logic programming language such as RELFUN with the multi-agent (MA) paradigm and the methodology coming from research in the computer supported cooperative work area. Thus we carefully selected a particular specialty within medicine, concerning the management of patients affected by acute myeloid leukemia (AML), which emphasizes the exploitation of different knowledge types coming from distinct professionals and stresses interoperability issues among them such as cooperation and negotiation.

AML is a clonal, malignant disease of the hemopoietic tissue that is characterized by the proliferation of abnormal (leukemic) blast cells principally in the marrow and by an impaired production of normal blood cells (Lichtman, 1991). It usually originates from a single transformed stem cell controlling hemopoiesis which is capable of differentiation into all blood cell lineages. This leukemic cell rapidly propagates itself in an abnormally regulated and disordered fashion and eventually will end up replacing all the normal blast cell population. AML therefore requires a prompt diagnosis and subsequent planning and administering of a strong therapy to eradicate the malignant cells, as any delay may turn out to have negative influence on the patient's life expectancy. The process of diagnosing AML usually starts from some general and very unspecific signs and symptoms and goes on through more specific laboratory exams. Finally, in

Figure 1: The task planning diagram for acute myeloid leukemia (AML)

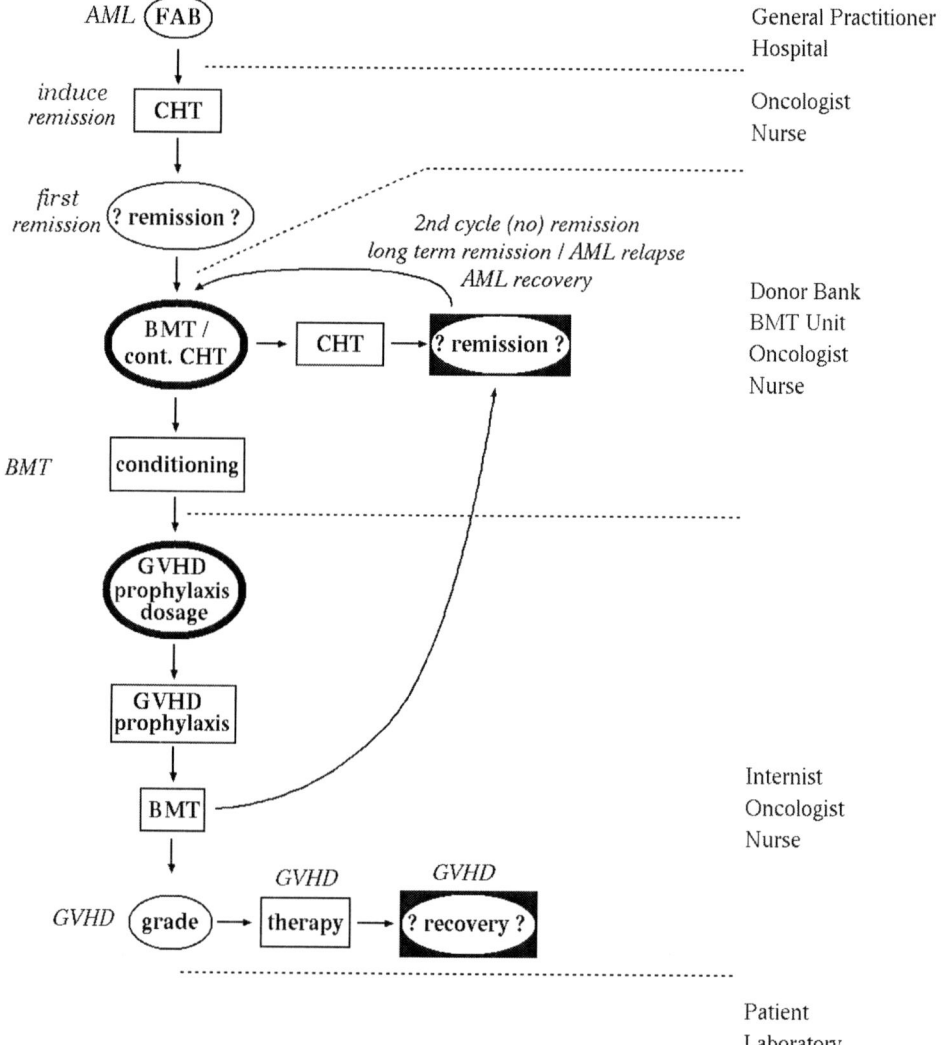

view of planning a suitable therapy, AML is ranked according to the French American British (FAB) classification into classes FAB-M1 through FAB-M7 (Bennet et al., 1985). Basically there are only two therapies available for AML, which are bone marrow transplantation (BMT) and chemotherapy (CHT). The former has the greatest probability of success, but it is only applicable in some specific situations which clearly include the availability of a compatible donor. It also has a major counter-indication mainly consisting in the possibility for the patient to develop graft-vs-host disease (GVHD).

To summarize the AML management protocol, Figure 1 gives an AML task planning diagram where all the different actions are shown, pointing out how they relate to each other. Ovals represent decision steps in the management process and fall into two different types. Thin ovals refer to diagnostic actions aimed at assessing the state of the patient, while thick ones represent therapy planning stages useful for selecting the most appropriate treatment based on the patient data and on the currently available diagnosis. Rectangles simply refer to therapeutic actions which are usually performed according to a given protocol and do not require any specific decision-making process. Finally, ovals encapsulated in black rectangles represent a monitoring action. They represent a process aimed at continuously assessing the patient state over time, looking for the occurrence of some particular events.

In the rightmost part of Figure 1 the main actors involved in each step of the management process are also indicated. The patient and the laboratory are supposed to be involved in each step, so they are indicated just once for all at the bottom. At the beginning of the management process the only additional actors will be the general practitioner (GP) and the hospital. After getting the results from the laboratory and diagnosing AML, the GP must interact with some hospitals in order to hospitalize the patient, and depending on AML seriousness and the availability of beds, this may require some negotiation. More specifically, the GP may wish to trade time with the possibility of selecting the hospital. If there are no available beds at the nearby hospital, depending on AML seriousness, the GP may wish to wait for one to become vacant or move the patient to a more distant hospital. Then, after a first CHT cycle is finished, the main therapy for the patient must be selected. This step requires a careful cooperation among the oncologist, the donor bank and the BMT unit. In fact, selecting the right therapy not only depends on the patient findings alone, but also on the possibility of finding a suitable donor. So, even if in principle BMT may be the most appropriate treatment, its suitability must be weighed against the availability of a suitable donor given by the donor bank and compared to that of CHT, as suggested by the oncologist. Moreover, the availability of a suitable donor may be a time dependent function, adding extra complexity to the problem.

As it transpires from the simplified description above, the whole process of managing AML-affected patients is a knowledge intensive activity involving several healthcare professionals who must be able to exchange information and use different reasoning techniques such as qualitative models (Ironi et al., 1990), or probabilistic reasoning (Quaglini et al., 1994).

With the aim of providing a specialized tool helping every single professional involved in AML management, we implemented a set of agents, each one encapsulating the domain knowledge of a given professional. All those agents were also endowed with cooperation knowledge, in order that they be able to interoperate and perform negotiations among each other on the basis of the same model adopted by human agents. In the next section we shall provide a description of how RELFUN, complemented with a KQML communication protocol for message exchanging, was used for this purpose.

THE FUNCTIONAL-LOGIC IMPLEMENTATION

The implementation of our prototypical system has basically involved two different parts. The first one addresses the modeling of domain specific knowledge within each agent being implemented through RELFUN. To support this, a frame-based representation of medical knowledge has been worked out, and deduction and abduction rules have been subsequently implemented for each specific medical task. The second part concerns modeling and representing cooperation knowledge in terms of a suitable set of rules implemented within each agent and aimed at achieving their ability to interoperate. This is based on the KQML multi-agent environment described in a previous section. The following subsections will explain both parts with typical examples; the full implementation is available and described in Campagnoli (1996).

Representing Domain Knowledge

RELFUN has been very helpful in representing structured knowledge within each single agent, providing an easy implementation of an inheritance mechanism in a frame-based (OOP) hierarchy.

For example, the diagnoser agent has the task of initially assessing the patient state, coming up with a solution which is then used as a starting point by those devoted to planning a therapy. Diseases are ranked in a taxonomy, where a class-subclass link is the leading one for arranging them according to pathophysiological criteria. The following KB fraction shows how `acute-leukemia` is defined as a `leukemia`, having acute myeloid leukemia (`aml`) as its subclass. The latter has instances such as `fab-m1, fab-m2, ...` (actually through `fab-m7`), representing the actual disease type.

```
class(leukemia,top,[]).
class(acute-leukemia,leukemia,
[[examinations,bone-marrow-blasts]]).
class(aml,acute-leukemia,
[[examinations,[blast-pas,black-sudan]]]).
instance(fab-m1,aml,
[[treatments,[bmt,cht]],
[examinations,blasts-morphology]]).
instance(fab-m2,aml,
[[treatments,[bmt,cht]]]).
...
```

Each disease class also has attributes, and the only one shown in the example above is that concerning the `examinations` required to find out if the patient is affected by the corresponding disease. Given the nature of the problem it is not possible to reach a satisfactory diagnostic conclusion at a high level in the taxonomy, as its determination depends on very specific findings which are usually associated with low-level subclasses and disease instances. To this end we have to implement a specific inheritance mechanism which calls upon discarding superclass values in presence of any local value for the `examinations` attribute. The KB fraction performing this task is shown below:

```
override(Slots,[]) :& Slots.
override(Specific,[[Attribute,_]|General]) :-
    member([Attribute,_],Specific)
    & override(Specific,General).
override(Specific,[[Attribute,Value]|General]) :-
    naf(member([Attribute,_],Specific))
    & tup([Attribute,Value]|override(Specific,General)).
inherit(top,Slots) :& Slots.

inherit(Class,SpecificSlots) :-
    class(Class,Superclass,ClassSlots)
    &
    inherit(Superclass,override(SpecificSlots,ClassSlots)).
slots(Instance) :-
    instance(Instance,Class,InstanceSlots)
    & inherit(Class,InstanceSlots).
```

Representing Cooperation Knowledge

Our effort is now focused on the implementation of the several agents involved in the distributed architecture emulating the interoperational model adopted by humans. Each agent, after being endowed with the domain specific knowledge corresponding to a given healthcare professional, is therefore also being provided with an explicit model of the strategies adopted by that professional for interacting with others.

As a simple example describing how the interoperation among different agents has been achieved, let's specify the start of the AML task planning in section 4, namely the strategy adopted by the medical agent emulating the general practitioner (GP). The GP's first task, as sketched in the AML task planning diagram in Figure 1, will consist in assessing the patient state in search for a diagnosis. Once this has been accomplished and the diagnosis has been identified as AML, the next task to be carried out by GP will deal with hospitalizing the patient. The following clauses, stored within the GP knowledge base, characterize the seriousness of the conditions of two patients who have been diagnosed as affected by AML and list three hospitals, ordered according to their distance from the patient's dwelling:

```
seriousness(patient1)  :&  stage3.
seriousness(patient2)  :&  stage5.
hospital(h1).
hospital(h2).
hospital(h3).
```

Then GP will send out a KQML message to a facilitator (see section 3) looking for some information from those hospitals concerning bed availability and the associated costs.

```
send-msg(gp,facilitator,ask-all,
    [Result,availability[Month]],_,a1).
```

The facilitator, after contacting all the available hospitals in behalf of GP, will provide a summary of the gathered [*hospital,cost*] `availability` results, sending it back to GP as a reply to the original query. What is returned will be stored as a set of clauses within GP's knowledge base and will look like the following (the non-deterministic `availability` function represents the facilitator's multiple results from the hospital agents for each fixed month):

```
availability(january)  :&  [h1,900].
availability(january)  :&  [h2,750].
availability(january)  :&  [h3,300].
availability(february)  :&  [h1,800].
availability(february)  :&  [h2,450].
availability(february)  :&  [h3,200].
. . .
```

Based on that information and on a suitable selection strategy which accounts for the hospital distance, charged costs and bed availability as a function of time, and the possibility to delay the patient hospitalization according to the seriousness of its state, GP will choose a hospital:

```
choose(Date,Patient)  :-
    hospital(Hospital),
    [Hospital,Cost] .= availability(Date),
    acceptable(Cost,seriousness(Patient)),
    confirm(Hospital,Patient)
    & Hospital.
```

If the previous rule fails, which means that it has not been possible to identify a solution according to the strategy it models, the following rule pair will be applied:

```
choose(Date,Patient)  :-
    stage5 .= seriousness(Patient),
    hospital(Hospital),
    [Hospital,_] .= availability(Date),
    confirm(Hospital,Patient)
    & Hospital.
choose(_,Patient)  :-
```

```
naf(stage5 .= seriousness(Patient)),
[Hospital,_] .= min-cost(tupof(availability(_))),
confirm(Hospital,Patient)
& Hospital.
```

The first one chooses the nearest available hospital without taking into account any cost if the patient condition is such that it requires an immediate treatment. The second one, which will be invoked when the patient's condition is less deteriorated, selects the cheapest hospital available over the whole year.

For implementing the hospital selection strategy, the cost charged by the hospital is compared to the seriousness of the patient as expressed by the AML stage according to the following rules:

```
acceptable(Cost,Seriousness) :-
    Seriousness .= cost(Min,Max),
    <=(Min,Cost),
    >=(Max,Cost).
cost(0,100):& dom[stage1,stage2,stage3,stage4,stage5].
cost(100,200) :& dom[stage3,stage4,stage5].
cost(200,300) :& dom[stage4,stage5].
cost(300,1000) :& dom[stage5].
```

Finally, in the above-mentioned cases, when the strategy has succeeded, the `choose` rule will also take care of contacting the hospital for issuing a reservation for the patient. This is accomplished through the rule:

```
confirm(Hospital,Patient) :-
    send-msg(gp,Hospital,tell,[confirm,Patient],_,ack1).
```

CONCLUSIONS

If the functional-logic paradigm wants to get closer to mainstream languages, as accomplished by the relational-database (SQL3) and OOP (Smalltalk, Java, and C++) paradigms, it, too, has to tackle real-world applications and handle their non-declarative aspects. The multi-agent medical application described in this paper has confirmed the practical use of purely functional-logic features such as non-ground function calls for function inversion. Even the RELFUN implementation of overriding inheritance is pure, and the entire prototype does not yet require a cut operator. On the other hand, we have found problems with the lack of LISP- or PROLOG-like KB-update operators such as assert and retract in modern functional-logic languages: the RELFUN users needed a dynamic assert because they wanted KQML-content parts of messages to become available as clauses in a receiving agent–clauses not different from those pre-existing in the receiver's KB, hence not somehow "coded" as data; conversely, users needed retract. In a delivery version of the prototype, RELFUN's module system could be used to 'localize'–and buffer–message I/O.

After some initial reservations by the RELFUN developers, the needed non-declarative extensions were made. Interestingly, functional-logic integration offered advantages compared to functional or logic programming alone even for these non-declarative primitives: Both assertz and retractx are value-returning, permitting assertz-modify-retractx compositions; retractx itself need only retract one eXactly specified clause, since for unification-based retractions, clause generation can be reused from RELFUN's clause-returning built-in via non-deterministic retractx-clause compositions. Also, in functional-logic languages, such non-declarative primitives can be built on top of an already integrated declarative kernel, thus avoiding their duplication in the two component languages. The same principle was already successful for declarative, not integration-relevant but uniformity-preserving extensions of the pure RELFUN kernel such as finite domains, which are relegated to outer pure RELFUN shells.

Summarizing the experience with our prototype, besides the assertz/retractx extension by the developers, the current RELFUN interpreter turned out to be easy to extend and use: For example, the users were able to implement the KQML interface via LISP built-in extensions with little help from the developers. Because of its KB hashing, the interpreter was also efficient enough for this communication-intensive application. With the RELFUN compiler plus (C) emulator (Perling, 1998), this functional-logic prototype could be developed into a real-world application.

While the prototype exploits several specifics of RELFUN, it is fairly clear that other implemented functional-logic integrations such as ALF and BABEL could have been used instead, adapting things to their language concepts. For example, our current frame-based hierarchy accesses *lists* of attribute-value pairs recursively, which could be built in via the notion of *feature terms* (Smolka & Treinen, 1992), as, e.g., implemented in Oz, and also envisaged as a RELFUN extension. For functional-logic comparison, it would thus be interesting to reimplement (parts of) this architecture in these other languages and, for ultimate reuse, in a standard proposal like Curry (Hanus et al., 1995).

With the advent of XML, the functional-logic rules of our prototype could now also be represented in markup languages such as RFML (http://www.relfun.org/rfml) or the standard proposal RuleML (http://www.dfki.de/ruleml). Besides such XML-encoded content, also the KQML performatives of agent messages could be marked up in XML and RDF; this is one of the topics of the current DFKI FRODO project (http://www.dfki.uni-kl.de/frodo).

The ever-increasing costs faced by the delivery of healthcare services, combined with the need to rapidly spread the latest reasearch achievements, require a better organization and structuring of the practice meant as a teamwork. To this aim the Pavia group is merging the agent methodologies with organizational issues and workflow modeling techniques towards the implementation of a distributed system supporting healthcare workers in planning and coordinating their activities (Lanzola et al., 1999).

ENDNOTE

1 This chapter is based to a large part on joint work with Massimiliano Campagnoli (Campagnoli 1996). The Pavia team also included Sabina Falasconi and Roberta Saracco. The Kaiserslautern team also included Simone Andel, Markus Perling, and Michael Sintek.

REFERENCES

Bennet, J. M., Catovsky, D. and Daniel, M. T. (1985). Prepared revised criteria for the classifiction of acute myeloid leukemia: A report of the French American British Cooperative Group. *American Journal of Medicine*, 103, 620-629.

Blois, M. (1984). *Information and Medicine: the Nature of Medical Descriptions*. Berkeley Press.

Boley, H. (1999). *A Tight, Practical Integration of Relations and Functions. Berlin,* Heidelberg: Springer-Verlag, LNAI 1712.

Boley, H., Buhrmann, U. and Kremer, K. (1994). Towards a sharable knowledge base on recyclable plastics. In McDowell, K. J. and Meltsner, J. K. (Eds.), *Knowledge-Based Applications in Material Science and Engineering*, TMS, 29-42.

Campagnoli, M. (1996). Una Architettura per la Comunicazione e Cooperazione fra Agenti Software. *Technical Report,* University of Pavia.

Finin, T., Weber, J., Wiederhold, G., Genesereth, M. R., Fritzson, R., McKay, D., McGuire, J., Pelavin, P., Shapiro, S. and Beck, C. (1992). Specification of the KQML Agent Communication Language. *Technical Report, EIT 92-04,* Enterprise Integration Technologies, Palo Alto, CA.

Fisher, M. and Wooldridge, M. (1993). Specifying and verifying distributed intelligent systems. In Filgueiras, M. and Damas, L. (Eds.), *Progress in Artificial Intelligence, Proceedings EPAI '93. Lecture Notes in Artificial Intelligence*, 727, 13-28), Springer-Verlag.

Hanus, M. (1994). The integration of functions into logic programming: From theory to practice. *Journal of Logic Programming*, 19(20), 583-628.

Hanus, M., Kuchen, H. and Moreno-Navarro, J. J. (1995). Curry: A truly functional logic language. In *Visions for the Future of Logic Programming: Laying the Foundations for a Modern Successor to Prolog. Proc. Workshop in Association with ILPS '95*, Portland, Oregon, 95-107.

Ironi, L., Stefanelli, M. and Lanzola, G. (1990). Qualitative models in medical diagnosis. *Artificial Intelligence in Medicine*, 2, 85-101.

Jennings, N. R., Varga, L. Z., Aarnts, R. P., Fuchs, J. and Skarek, P. (1993). Transforming standalone expert systems into a community of cooperating agents. *Engineering Applications in Artificial Intelligence*, 6(4), 317-331.

Klinker, G., Bhola, C., Dallemagne, G., Marques, D. and McDermott, J. (1991). Usable and reusable programming constructs. *Knowledge Acquisition*, 3, 117-135.

Lanzola, G. and Stefanelli, M. (1993). Inferential knowledge acquisition. *Artificial Intelligence in Medicine*, 5, 253-268.

Lanzola, G., Falasconi, S. and Stefanelli, M. (1996). Cooperating agents implementing distributed patient management. In Van de Velde, W. and Perram, J. (Eds.), *Agents Breaking Away. Proceedings MAAMAW '96*, Eindhoven, The Netherlands, Lecture Notes in Artifical Intelligence, 1038, 218-232, Springer-Verlag.

Lanzola, G., Gatti, L., Falasconi, S. and Stefanelli, M. (1999). A framework for building cooperative software agents in medical applications. *Artificial Intelligence in Medicine*, 16, 223-249.

Lenat, D. B. and Feigenbaum, E. A. (1991). On the treshold of knowledge. *Artificial Intelligence*, 47, 185-250.

Lichtman, M. A. (1991). Hemopoietic stem cell disorders: Classification and manifestations. *Hematology*, McGraw-Hill, 148-157.

Neches, R., Fikes, R. E., Finin, T., Gruber, T. R., Patil, R., Senator, T. and Swartout, W. (1991). Enabling technology for knowledge sharing. *AI Magazine*, 12, 36-56.

Perling, M. (1998). The RAWAM: Relfun-Adapted WAM Emulation in *C. Technical Memo TM-98-07*, DFKI GmbH, December.

Quaglini, S., Bellazzi, R., Locatelli, F., Stefanelli, M. and Salvaneschi C. (1994). An Influence Diagram for Assessing GVHD Prophylaxis After Bone Marrow Transplantation in Children. *Medical Decision Making*, 14, 223-235.

Ramoni, M., Stefanelli, M., Magnani, L. and Barosi, G. (1992). An epistemological framework for medical knowledge-based systems. *IEEE Transactions on Systems, Man, and Cybernetics*, 22(6), 1361-1375.

Searle, J. R. (1969). *Speech Acts: An Essay in the Philosophy of Language.* Cambridge University Press.

Shoham, Y. (1993). Agent Oriented Programming. *Artificial Intelligence*, 60, 51-92.

Smolka, G. and Treinen, R. (1992). Records for logic programming. *Proceedings Joint International Conference and Symposium on Logic Programming*, Washington, DC, MIT Press.

Chapter III

Cross-Fertilizing Logic Programming and XML for Knowledge Representation

Harold Boley
German Research Center for Artificial Intelligence, Germany

INTRODUCTION

The simplicity of Web-based data exchange is beneficial for nonformal, semiformal and formal documents. For formal specifications and programs the Web permits distributed development, usage and maintenance. Logic programming (LP) has the potential to serve as a uniform language for this. Meanwhile, however, the World Wide Web Consortium (W3C; http://www.w3.org/) has enhanced HTML–for nonformal and semiformal documents–into the Extensible Markup Language (XML) (Harold, 1999)–for semiformal and formal documents.

This raises the issue of the relationships between XML and LP. Will logic programming have the chance, despite, or perhaps precisely because of XML, to become a 'Web technology' for formal documents? Could the HTML-like syntax of XML be replaced by a Prolog-like syntax, or could it be edited or presented over a standardized stylesheet—in such a Prolog syntax? Is SLD resolution a suitable starting point for the interpreter semantics of an XML query language like XQL (http://www.w3.org/TandS/QL/QL98/pp/xql.html) or should an LP-oriented, inferential query language be developed in the form of an XML-based Prolog? In the following text, such questions will be discussed, and possible interplays between XML and LP—in both directions—will be presented.

The already foreseeable success of XML as a 'universal' interchange format for commercial practice and healthcare can also be viewed as a success of the declarative representation technique as proposed, in a somewhat different form, by logic programming. Similarities and differences between these declarative paradigms will be later elaborated upon. Let us here state two success factors of XML: (1) a sufficient compatibility of XML with the already widely-used HTML, e.g., through XHTML, and (2) the XML standardization through the W3C and, consequently, the building of XML tools into current browsers like Internet Explorer and Netscape/Mozilla and, hence, the increase of the number of XML documents in the Web. The ISO standard of Prolog (Deransart, Ed-Dbali & Cervoni, 1996) can hardly attain the same broad effect because of its missing Web/HTML orientation, even though it has a more strongly focused semantics. It would be interesting to compare the two standards in a more detailed manner than is possible within the limited scope of this paper. The path for (partial) Prolog-XML translators, interfaces, etc. could thus be paved in both directions, combining the (essential) advantages of both paradigms.

Beyond all start-up euphoria, XML offers new general possibilities from which logic programming can also benefit: (1) definition of self-describing data in an internationally standardized, non-proprietary format, (2) structured exchange of data and knowledge between enterprises of various industries, and (3) integration of information from various sources into unified documents. Furthermore, as already discussed in Boley (1996)—for the XML predecessor SGML—XML is the most suitable language for logical knowledge bases in the Web. Additional LP use of XML would be the exchange of knowledge bases between different logic languages as well as between LP knowledge bases and databases, application systems, etc. Even the transformation or compilation of logic source programs could be done on the basis of XML markups and annotations. These and similar possibilities will be expanded in subsequent sections.

The current text uses concrete comparisons of LP and XML examples in order to work out (syntactic) differences as well as (semantic) similarities. The text is directed primarily to readers who wish to understand basics of XML via LP and deductive or relational databases. But also in the converse direction, aspects of LP appear in a new light via XML. The sections after this one are constructed as follows: Next comes an introduction of 'elementary' XML and an element representation by means of Prolog structures. Thereafter we look at the reverse direction, demonstrating how Herbrand terms can be represented as XML elements. Building on that, we deal with an XML representation of Horn clauses (pure XmlLog). Subsequently, XML attributes like `id`/`idref` are employed for extended logics. After this, XML's document type definitions are used for logic languages like XmlLog. Then, XML query languages like XQL and inferential logic programs are compared. In conclusion, we discuss some similar approaches and future work.

XML ELEMENTS AS GROUND STRUCTURES

XML documents consist of (nested) elements. Each element is a sequence of the form *<tag>...</tag>*, i.e. some content '. . .' enclosed by *tag*-'colored' start- and end-brackets. This content may consist of text or again of elements. Consequently, a 'start-tag' and 'end-tag' can be used for the total markup of some textual content, which may contain an arbitrary amount of further well-bracketed detailed markup. With the increase of XML markup in a document its 'formality' will also increase in the sense of the spectrum mentioned at the beginning of the introductory section.

Let us next consider an **informal** minimal markup with a single tag pair
```
<sentence>...</sentence>:
<sentence> Medionbook sold 10358 copies of XML4Med on-line
</sentence>
```

Through rough parsing one obtains a **semiformal** `<triple>` element, whose `<predicate>` and `<object>` subelements contain unanalyzed text:
```
<triple>
    <subject> Medionbook </subject>
    <predicate> sold on-line </predicate>
    <object> 10358 copies of XML4Med </object>
</triple>
```

Finally, a complete, context-free parsing yields a (syntactically) **formal** document with the following `<s>`-rooted element nesting:
```
<s>
 <np>
    <noun> Medionbook </noun>
 </np>
 <vp>
   <vgroup>
    <verb> sold </verb>
    <adverb> on-line </adverb>
   </vgroup>
   <np>
    <ngroup>
      <card> 10358 </card>
      <noun> copies </noun>
    </ngroup>
    <pp>
     <prep> of </prep>
```

```
      <np>
        <noun> XML4Med </noun>
      </np>
    </pp>
   </np>
  </vp>
</s>
```

A document, therefore, is structured through XML in the form of a parse tree or a derivation tree, whereby each element <*tag*>...</*tag*> is a *tag*-marked node (a nonterminal) which has ordered, directed edges with direct subelements, appearing in '...' (shown here through indent). The texts appearing in '...' are the leaves of the derivation tree. In the normal case, each XML element contains only subelements (corresponding to nonterminals) or only text (corresponding to a 'token' of terminals).

Apparently, such an XML document is–in Prolog–comprehendable as a (variable-free, i.e.) ground structure *tag*(., ., .), provided that texts on the leaves can be represented as individual constants. Thereby *tag* becomes the constructor, and the contents '...' becomes correspondingly transformed arguments '., ., .'. Without going into the exact 'character escape'-conventions of XML and Prolog, we represent XML texts as Prolog strings; after complete analysis we use Prolog individual constants (with lowercase letters).

The three steps of parsing the example in Prolog therefore lead to the following ground structures:

```
sentence("Medionbook sold 10358 copies of XML4Med on-line")
triple(
     subject("Medionbook"),
     predicate("sold on-line"),
     object("10358 copies of XML4Med")
     )
s(
 np(noun(medionbook)),
 vp(
     vgroup(verb(sold),adverb(on-line)),
     np(ngroup(card(10358),noun(copies)),
        pp(prep(of),np(noun(xml4med))))
     )
 )
```

Our linguistic analysis now proceeds from this general syntax to special semantics, e.g., for the discourse domain 'e-health.' Analysts of specific discourse domains often jump directly to such semantics, without our previous discussion of the natural-language (-XML) part.

The result in XML is a (semantically) formal document with four <sales> subelements, corresponding to domains of a relational <sales> database table:

```
<sales>
    <channel> on-line </channel>
    <company> Medionbook </company>
    <item> XML4Med </item>
    <quantity> 10358 </quantity>
</sales>
```

This, then, in Prolog leads to the following ground structure:

```
sales(
    channel(on-line),
    company(medionbook),
    item(xml4med),
    quantity(10358)
    )
```

HERBRAND TERMS AS XML ELEMENTS

The representation of Herbrand terms (individual constants, logic variables, and structures) in XML requires, in essence, the use of <ind>, <var>, and <struc> elements.

Like all elements, an *individual constant* in XML must be explicitly labeled as such through markup, here an *<ind> element* of the form <ind>...</ind>. Thus the Prolog individual constant spinal-cord, for the spinal cord between the brain and the body, becomes the XML element <ind> spinal-cord </ind>.

In Prolog, instead of an individual symbol, a *(ground) structure* like neural-connection(brain,body) can also be used for the denotation of an individual. One can also view this as a 'record' with components for the connected organ systems. In XML, a structure becomes a *<struc> element* with an embedded element for the constructor, followed by elements for its argument terms, here two individual constants:

```
<struc>
    <constructor> neural-connection </constructor>
    <ind> brain </ind>
    <ind> body </ind>
</struc>
```

Because the XML markup makes the syntax of the original Prolog structure explicit, this can be regained from it as well as a Prolog structure can be constructed, whose syntax is described as follows:

```
struc(
    constructor(neural-connection),
    ind(brain),
    ind(body)
    )
```

If the original structure is already nested as, for example,

```
circulation-system(
 neural-connection(
    brain,
    organ-compound(
      thorax,
      abdomen,
      left-arm,
      right-arm,
      left-leg,
      right-leg)))
```

this, then, leads to embedded `<struc>` elements as in

```
<struc>
  <constructor> circulation-system </constructor>
  <struc>
   <constructor> neural-connection </constructor>
   <ind> brain </ind>
   <struc>
     <constructor> organ-compound </constructor>
     <ind> thorax </ind>
     <ind> abdomen </ind>
     <ind> left-arm </ind>
     <ind> right-arm </ind>
     <ind> left-leg </ind>
     <ind> right-leg </ind>
   </struc>
  </struc>
</struc>
```

By start-tag/end-tag bracketing, XML elements are always explicitly type-labeled. Therefore, in Prolog, one does not intuitively see from the symbols `spinal-cord` and `circulation-system` that the first is an individual constant and the second is a constructor. In XML, as in strongly typed LP languages, such distinctions are explicit, but XML shows them at every occurrence of a symbol. The example above demonstrates the self-description advantage–but also shows the 'space requirement'– of this generous application of 'syntactic sugar.' The redundancy in writing every closing XML bracket as a full 'end-tag' can be eliminated by the use of the 'neutral' closing bracket `</>`–approaching Prolog structures–as in XML-QL (http://www.w3.org/TR/ NOTE-xml-ql/). This, however, is not standardized in XML 1.0.

Lists–as they are, e.g, used in Prolog–can also be reduced to nested `cons`- structures or be represented as n-tuples (Boley, 1999) in XML directly.

In XML, a *logic variable* is marked through a *<var> element* of the form `<var>...</var>`. Thus the Prolog variable `Xyz` becomes the XML element `<var> xyz </var>`, whereby the `<var>` markup makes conventions like first-letter capitalizing superfluous.

Hence a (not variable-free, i.e.) 'non-ground' structure like Prolog's `neural-connection(brain,Xyz)` becomes XML's non-ground element

```
<struc>
    <constructor> neural-connection </constructor>
    <ind> brain </ind>
    <var> xyz </var>
</struc>
```

HORN CLAUSES AS XML ELEMENTS

The representation of Horn clauses (facts and rules) in XML calls for further elements. A predicate or *relation symbol* will in XML be a *<relator> element*. For example, the relation symbol `travel` will be `<relator> travel </relator>`.

The *application of a relation symbol to terms* is marked by a *<relationship> element* in XML. A `travel` application `travel(knee-jerk-reflex,spinal-cord)` to two individual constants will thus be

```
<relationship>
    <relator> travel </relator>
    <ind> knee-jerk-reflex </ind>
    <ind> spinal-cord </ind>
</relationship>
```

Moreover, a Horn fact in XML is asserted as an *<hn> element* that possesses exactly one subelement–the *<relationship> element*. In the example, the Prolog fact

`travel(knee-jerk-reflex,spinal-cord).`

becomes

```
<hn>
    <relationship>
      <relator> travel </relator>
      <ind> knee-jerk-reflex </ind>
      <ind> spinal-cord </ind>
    </relationship>
</hn>
```

No further XML element types are required to represent Horn rules. A *relation call* is written again as a *<relationship> element*. The `carry` call `carry(vertebral-column,Pulse)`, with an individual constant and a logic variable, becomes

```
<relationship>
    <relator> carry </relator>
    <ind> vertebral-column </ind>
    <var> pulse </var>
</relationship>
```

A Horn rule, then, in XML is asserted as an *<hn> element* that has two or more subelements, the head-`<relationship>` element followed by at least one body-`<relationship>` element. The above example is thus generalized to the Prolog rule

```
travel(Pulse,spinal-cord) :-
    carry(vertebral-column,Pulse).
```

and rewritten in XML as

```
<hn>
    <relationship>
      <relator> travel </relator>
      <var> pulse </var>
      <ind> spinal-cord </ind>
    </relationship>
    <relationship>
      <relator> carry </relator>
      <ind> vertebral-column </ind>
      <var> pulse </var>
    </relationship>
</hn>
```

Pure XmlLog–the pure Prolog in XML syntax that was up to now introduced by examples–will be precisely defined in a later section.

ATTRIBUTES FOR EXTENDED LOGICS

While nested elements–as trees–can represent arbitrary information, in some situations this leads to unnecessarily deep- or broad-nested representations. In XML, remedies are often obtained through *attributes*: A start-tag can be enhanced by n attribute-value pairs of the form $a_i = v_i$; in general, every element thus is a sequence of the form $<tag \ a_1 = v_1 ... a_n = v_n> \ ... \ </tag>$.

For example, the `<sales>` element at the end of the second section can be written more compactly and intuitively through the conversion of the `<channel>` subelement into a `channel` attribute with the string value "`on-line`":

```
<sales channel="on-line">
      <company> Medionbook </company>
      <item> XML4Med </item>
      <quantity> 10358 </quantity>
</sales>
```

While there is no direct counterpart to this in Prolog, other logic programming languages allow a corresponding conversion of the original Prolog structure. In the functional-logic Relfun (http://www.relfun.org/), one would obtain–apart from its use of square brackets-the structure

```
sales(channel(on-line))(
      company(medionbook),
```

```
item(xml4med),
quantity(10358)
                        )
```

with a constructor `sales(channel(on-line))` parametrized by the attribute.

Additional information is often also better directly introduced as attributes instead of as subelements. The `<noun> copies </noun>` used earlier, for example, is extensible, through attributes for gender and number, to `<noun genus="neutrum" numerus="plural"> copies </noun>`, respectively to `noun(genus(neutrum),numerus(plural)) (copies)`. Generally, this allows feature grammars to be reconstructed in XML and LP.

Independent of grammars, XML attributes can be used in logic programming for the representation of feature/psi terms. However, since the values in $a_i=v_i$ are atomic in XML 1.0, only feature/psi terms with corresponding atomic values can be directly represented. But this is not a problem for, e.g., description logics (Calvanese, De Giacomo & Lenzerini, 1999).

A particularly useful application of XML attributes for Prolog is the 'arity' labeling of relation symbols. Thus, in the pervious section, a relation symbol labeled as binary `travel/2` could be written in XML with an arity attribute as `<relator arity="2"> travel </relator>`. The same applies to arity labelings of constructors and–in functional-logic languages–of (defined) function symbols.

Similarly, attributes enable annotations at any element level, including mode declarations for logic variables, determinism specifications for clauses or procedures, and context conditions for entire knowledge bases.

Further, annotations which are necessary for compiling logic source programs can themselves be represented in a declarative intermediary language, e.g., in the form of 'classified clauses' (Krause, 1990). These can now be represented as attributed XML elements.

In addition to the previously used attributes with values of the general type `CDATA` ('character data'), XML permits attributes of more specific type, too. In particular, the types `ID` and `IDREF` exist for naming and referencing elements of a given XML document. An `ID`-typed value must uniquely identify some element, and an `IDREF`-typed value must also occur as an `ID`-value of some element.

With these XML constructs, many meta-logical and modal-logical assertions can be expressed. For example, clauses can be named and their names can be used for the construction of further clauses. In this section, we assume, as in many XML tools, the attributes `id` and `idref` are of the type `ID` and `IDREF`, respectively.

In such an extension of our pure XmlLog, we can name the Prolog fact of the previous section, for example, and store it in a knowledge-base module of 'hypothetical Prolog facts' as follows:

```
<hn id="knee-spinal">
    <relationship>
      <relator> travel </relator>
      <ind> knee-jerk-reflex </ind>
      <ind> spinal-cord </ind>
    </relationship>
</hn>
```

For the representation of the 'modal Prolog fact'
`belief(mary,travel(knee-jerk-reflex,spinal-cord)).`
we can access the `"knee-spinal"` proposition from another knowledge-base module as follows:

```
<hn>
    <relationship>
      <relator> belief </relator>
      <ind> mary </ind>
      <prop idref="knee-spinal"/>
    </relationship>
</hn>
```

Here, the XML abbreviation of empty elements *<tag ...> </tag>* to *<tag ...>/>* is used in order to concisely write the propositional argument of the `belief` operator as `<prop idref="knee-spinal"/>`. Note that the above Prolog-like notation for propositions as arguments–without a declaration of modal operators such as `belief` or a separation of relation symbols and constructors–is not distinguishable from that of structures as arguments. In the self-descriptive XML notation, the tags for `prop` and `struc` are clearly differentiated–regardless of the fact that we have represented propositions in XML only by their names.

XML's tree structure can be generalized for directed acyclic graphs (DAGs) using `ID/IDREF`. For example, a `disbelief` fact for `lydia` with `idref` can access the same `"knee-spinal"` proposition, i.e., `ID/IDREF` enable sharing. Moreover, one can also proceed to general directed graphs, which can, e.g., be used to represent regular trees in Prolog II+ (Colmerauer, 1984).

DOCUMENT TYPE DEFINITION FOR LOGIC LANGUAGES

In the previous sections we have introduced a Horn logic language, pure XmlLog, based on examples. XML also permits the general definition of this language.

In order to define languages syntactically, XML uses *document type definitions (DTDs)*, which we consider initially only as *ELEMENT declarations* for **non-attributed elements**. For the nonterminals, a DTD corresponds to an ordinary context-free grammar in a modified (EBNF) notation. For terminals, however, arbitrary permutations of the base alphabet are normally used (so-called 'Parsed Character Data,' or 'PCDATA'), instead of fixed terminal sequences. The abbreviation #PCDATA used in DTDs can also be considered as a nonterminal, which (as the Kleene closure) expands to all 'tokens' that can be generated with the alphabet.

With a DTD grammar, one can thus derive context-free word patterns with arbitrary 'tokens' as their leaves. This only makes sense, however, if the derivation tree itself–in a form linearized through brackets–is seen as the generated result. In XML this is achieved via 'keeping' the well-known brackets <*tag*>. . . </*tag*> for every nonterminal *tag*.

A given XML element is said to be *valid* w.r.t. a DTD, if it can be generated from the DTD as a linearized derivation tree in this way.

As an example of a DTD we consider the syntactic ELEMENT declarations for our pure XmlLog as a knowledge base (kb) of zero or more Horn clauses (hn*):

```
<!ELEMENT kb              (hn*) >
<!ELEMENT hn              (relationship, relationship*) >
<!ELEMENT relationship    (relator, (ind | var | struc)*) >
<!ELEMENT struc           (constructor, (ind | var | struc)*) >
<!ELEMENT relator         (#PCDATA) >
<!ELEMENT constructor     (#PCDATA) >
<!ELEMENT ind             (#PCDATA) >
<!ELEMENT var             (#PCDATA) >
```

With this DTD we show a generation of the XML-linearized derivation tree for the kb-embedded example rule in a previous section, which–as by an XML validator–can also be read from bottom to top as a parsing. Notice, for example, how the start symbol kb continues to bracket the clause(s) derived from it as part of the linearized start-tag/end-tag tree representation <kb>. . .</kb>. The omitted intermediate steps are easily completed:

```
kb
<kb> hn* </kb>
<kb> hn </kb>
<kb> <hn> relationship relationship* </hn> </kb>
<kb> <hn> relationship relationship </hn> </kb>
...
<kb>
 <hn>
  <relationship>
```

```
    <relator> #PCDATA </relator>
    <var> #PCDATA </var>
    <ind> #PCDATA </ind>
  </relationship>
  <relationship>
    <relator> #PCDATA </relator>
    <ind> #PCDATA </ind>
    <var> #PCDATA </var>
  </relationship>
 </hn>
</kb>

<kb>
 <hn>
  <relationship>
   <relator> travel </relator>
   <var> pulse </var>
  <ind> spinal-cord </ind>
  </relationship>
  <relationship>
   <relator> carry </relator>
   <ind> vertebral-column </ind>
   <var> pulse </var>
  </relationship>
 </hn>
</kb>
```

We now consider DTDs for **attributed elements**. For that we use *ATTLIST declarations*, which associate an element name with an attribute name plus its attibute type and a possible occurrence indication.

In a first example, we declare the `relator` attribute `arity` used earlier, as `CDATA`-typed (analogously to `#PCDATA`) and as occurring optionally (`#IMPLIED`):

```
<!ATTLIST relator  arity CDATA #IMPLIED >
```

As a second example we define the extended XmlLog introduced earlier with named `hn` clauses and embedded propositions. We first extend the above `ELEMENT` declaration of `relationship` by `prop` and declare the latter as an empty element:

```
<!ELEMENT relationship (relator, (ind | var | struc | prop)*) >
<!ELEMENT prop EMPTY >
```

Then we append `ATTLIST` declarations that specify for the `hn` and `prop` elements, respectively, the attribute `id`–as optional `ID` type–and `idref`–as mandatory `IDREF` type:

```
<!ATTLIST hn id ID #IMPLIED >
<!ATTLIST prop idref IDREF #REQUIRED >
```

Now, the previously named "knee-spinal" fact from the previous section can be generated/parsed together with its accessing facts using the complete DTD.

XML QUERY LANGUAGES AND LOGIC PROGRAMMING

The W3C standardization of XML query languages is not yet as far advanced as the W3C standardization of XML itself. Nevertheless, a preliminary comparison with logic programming is already possible and is addressed here on the basis of XQL (http://www.w3.org/TandS/QL/QL98/pp/xql.html) and XmlLog. Again, we consider the relationships in both directions.

Our **LP treatment of XQL queries** uses the document with the <s>-rooted element nesting from the second section. This document is extended by attributes for gender and number corresponding to a previous section. We assume that the document is accessible via a variable called Root.

The XQL query *//tag* returns all elements of the form *<tag ...>...</tag>* in the document. In the example, //np delivers the following three elements, where the third is contained within the second:

```
<np>
 <noun genus="neutrum" numerus="singular"> Medionbook </noun>
</np>


<np>
 <ngroup>
   <card> 10358 </card>
   <noun genus="neutrum" numerus="plural"> copies </noun>
 </ngroup>
 <pp>
  <prep> of </prep>
  <np>
    <noun genus="neutrum" numerus="singular"> XML4Med
    </noun>
  </np>
 </pp>
</np>
<np>
 <noun genus="neutrum" numerus="singular"> XML4Med </noun>
</np>
```

In pure logic programming, this behavior would be performed with a relation call descendant(Root,np,Res), which binds a result variable non-deterministically. Closer to the value-oriented thinking of XQL, a function call descendant(Root,np) can be used–say, in a functional-logic language–that returns the following Prolog structures as non-deterministically enumerated values:

```
np(noun(genus(neutrum),numerus(singular))(medionbook))
```

```
np(ngroup(card(10358),noun(genus(neutrum),numerus(plural))(copies)),
    pp(prep(of),np(noun(genus(neutrum),numerus(singular))(xml4med))))
```

```
np(noun(genus(neutrum),numerus(singular))(xml4med))
```

Through the XQL query *//tag[@a=v]* all elements of the form *<tag ...a=v...> ... </tag>* which contain the attribute-value pair *a=v* in the document are returned. In the example, `//noun[@numerus="plural"]` delivers only the following element:

```
<noun genus="neutrum" numerus="plural"> copies </noun>
```

The corresponding function call `descendant-filter (numerus(plural))(Root,noun)`, parameterized through a filter, returns the following Prolog structure:

```
noun(genus(neutrum),numerus(plural))(copies)
```

The definition of query functions like `descendant` and `descendant-filter( )` cannot, of course, consist only of (deterministic!) Prolog unifications: All Prolog substructures with the given constructors must be found via a recursive traversal, and filters must be applied to the constructor parameters.

Conversly, unification permits logic variables not only in the query, but also in the document, such as in the element for the non-ground structure `neural-connection(brain,Xyz)` from a previous section. In this way also fully unifying queries such as `neural-connection(Same,Same)`, which binds `Same` to `brain`, could be made on documents. The stored and universally readable non-ground element records, among others things, the idea that a brain part has a neural connection to another one. Non-ground documents can act as forms with some fields to be filled out identically; experience will reveal the extent to which such non-ground documents can be used in practice.

Let us turn now to **queries and inferences on the basis of the pure XmlLog** defined earlier.

Consider the fact `carry(vertebral-column,achilles-jerk)`. It can be used in its XmlLog version

```
<hn>
 <relationship>
  <relator> carry </relator>
  <ind> vertebral-column </ind>
  <ind> achilles-jerk </ind>
 </relationship>
</hn>
```

to answer the XmlLog version of the query `carry(vertebral-column,Pulse)` shown earlier, which binds the XML-logic variable `<var> pulse </var>` to `<ind> achilles-jerk </ind>`.

With such `carry` base facts, a rule can be applied that dynamically derives `travel` assertions on demand; their static storage is no longer necessary:

```
travel(Pulse,spinal-cord) :-
    carry(vertebral-column,Pulse).
```

Its XML version from a previous section is useful for inferential queries like `travel(achilles-jerk,Where)`:

```
<relationship>
 <relator> travel </relator>
 <ind> achilles-jerk </ind>
 <var> where </var>
</relationship>
```

This XML-version query unifies `<var> pulse </var>` = `<ind> achilles-jerk </ind>` and `<var> where </var>` = `<ind> spinal-cord </ind>` with the head of the rule. The internal XML query corresponding to `carry(vertebral-column, achilles-jerk)` that results with the first binding from the body of the rule succeeds with the above XML fact. Hence, the second binding, i.e., `Where = spinal-cord`, is supplied as the result of the external XML query. The entire inference is thus carried out as an SLD-resolution step (Lloyd, 1987).

In general, the procedural semantics of *SLD resolution* can be transferred to XmlLog. If, for this, we assume that the clauses are distributed over various documents in the Web, then we find ourselves in the situation described in Boley (1996): In this 'open world,' logical completeness, in particular, can no longer be expected. The (Java) realization of the semantics in the form of an XmlLog interpreter could occur on the client side as a standard extension–or plug-in–of a Web browser.

RELATED WORK AND CONCLUSIONS

The pure XmlLog introduced here is a sublanguage of **RFML** (Relational-Functional Markup Language; Boley, 1999), which also admits, among others, user-defined functions and higher-order operators (http://www.relfun.org/rfml/). **RuleML** (Rule Markup Language) attempts to standardize such backward rules as well as forward rules (http://www.dfki.de/ruleml).

SHOE (Simple HTML Ontology Extensions; Heflin, Hendler, & Luke, 1999) is the first well-known language that provides Horn clauses in the Web. It was initially specified–before XML was defined–in SGML and, meanwhile, also as an XML DTD (http://www.cs.umd.edu/projects/plus/SHOE/).

An interesting aspect of SHOE is that individual constants are represented through the (official) URL/URI for the individual. In our earlier Horn rule example, there appear two individuals, which could be represented as `vertebral-column=http://www.spine.org/` and `spinal-cord=http://www.spinalcord.org/`. In this way, the rule in SHOE becomes

```
<DEF-INFERENCE DESCRIPTION=
"travel(?pulse,http://www.spinalcord.org/) if
 carry(http://www.spine.org/,?pulse)">
 <INF-IF>
   <RELATION NAME="carry">
     <ARG POS="1" VALUE="http://www.spine.org/"/>
     <ARG POS="2" VALUE="pulse" USAGE="VAR"/>
   </RELATION>
 </INF-IF>
 <INF-THEN>
   <RELATION NAME="travel">
     <ARG POS="1" VALUE="pulse" USAGE="VAR"/>
     <ARG POS="2" VALUE="http://www.spinalcord.org/"/>
   </RELATION>
 </INF-THEN>
</DEF-INFERENCE>
```

'Courteous Logic Programs' (Grosof, Labrou, & Chan, 1999) have been studied in regard to the prioritized conflict treatment in agent communication for e-commerce (e.g., in supply contracts). For knowledge exchange an XML-based Business Rules Markup Language (**BRML**) has been developed (http://www.oasis-open.org/cover/brml.html), which also encodes the clause subset of the Knowledge Interchange Format (**KIF**) (http://logic.stanford.edu/kif/dpans.html) in XML.

Of the many suggested XML query languages, (http://www.w3.org/TandS/QL/QL98/) until now only **XQL** (http://www.w3.org/TandS/QL/QL98/pp/xql.html) and **XML-QL** (http://www.w3.org/TR/NOTE-xml-ql/) have been mentioned here.

A user-oriented alternative is **EquiX**, in which result documents with their DTDs are automatically constructed from queries over a form-based interface (Cohen et al., 1999).

The **algebra for XML query languages** (http://www.cs.bell-labs.com/who/wadler/papers/xquery-algebra/xquery-algebra.html) is implemented in the functional Haskell; reimplemented in the Haskell-generalizing functional-logic Curry (Hanus, 1997), it should be transferrable to logic programming. **An XML Query Algebra** now also provides the semantic basis of the wide-spectrum XML query language, **XQuery** (http://www.w3.org/TR/xquery).

More closely related to term-replacement systems than to logic programming is XSL (Extensible Stylesheet Language): For (restricted) term replacements, the XSL part **XSLT** (XSL Transformations; http://www.w3.org/TR/xslt) is used on XML documents; for this, XSLT again uses an XQL-like pattern language. A more exact characterization of the subclass of XSLT term-replacement systems (http://www.wi.leidenuniv.nl/home/maneth/DOOD2000.ps.gz) would be interesting in respect to the representation of XML elements as ground terms in this paper. The corresponding processing of non-ground terms with narrowing has been examined in functional-logic languages like Curry and Relfun.

Limited scope does not permit the **XML namespaces** to be discussed here, although interesting relationships with the module systems of LP languages do exist. These are generally accepted as the more difficult part in the ISO standardization of Prolog (http://www.logic-programming.org/prolog_std.html); on the other hand, they are not totally specific for logic programming.

The XML extensions on top of namespaces, **RDF** for the description of metadata and **RDF Schema** for the corresponding schema/ontology specification are also beyond the scope of this paper. RDF queries and RDF inferences, however, are subjects for techniques from the field of logic programming such as the application of F-Logic (Decker, Brickley, Saarela, & Angele, 1998). A variant of Protégé-2000 is suitable as an RDF editor (http://smi-Web.stanford.edu/projects/protege/protege-rdf/protege-rdf.html).

In this context, it should be noted that XML 1.0 does not offer the usual built-in types like `integer`, `float`, etc. Therefore, in a previous section, for example, we had to type the attribute `arity` with unspecific `CDATA` strings, although only the natural numbers are involved. This is where **XML Schema** comes to the rescue with its differentiated type system in part 2 (http://www.w3.org/TR/xmlschema-2/). Further studies could investigate the relationships of the type/sort or signature systems of (feature/psi-term-)LP languages to those of XML Schema and RDF Schema.

In summary, XML presents–extending the Internet workshops at LP conferences since 1996–new areas of analysis and application for the field of logic programming. Both areas benefit from the interactions discussed. Several of the mentioned XML extensions are still in the preliminary draft stage and are not yet ready for (W3C) standardization. Therefore, it might still be possible to have an influence on the final form of these XML extensions.

ACKNOWLEDGMENTS

I would like to thank Ulrich Geske and the program committee of the 14th Workshop on Logic Programming for the invitation. For the proofreading of a first version my thanks go to Andreas Klüter and Michael Sintek. The English translation was supported by Christy Anderson, Brice Kamga, Kushalappa Paleyanda, and Glenn Peach. Special thanks go to Rolf Grütter for his valuable questions and suggestions as well as his editorial patience and help.

REFERENCES

Boley, H. (1996). Knowledge Bases in the World Wide Web: A Challenge for Logic Programming. In Tarau, P., Davison, A., De Bosschere, K. and Hermenegildo, M. (Eds.), *Proc. JICSLP '96 Post-Conference Workshop on Logic Programming Tools for INTERNET Applications*, 139-147. COMPULOG-NET, Bonn. Revised versions in International Workshop "Intelligent Information Integration," KI-97, Freiburg, September, 1997. DFKI *Technical Memo TM-96-02*, October, 1997.

Boley, H. (1999). The Relational-Functional Markup Language RFML. *Technical report, Stanford Medical Informatics*, December.

Calvanese, D., De Giacomo, G. and Lenzerini, M. (1999). Representing and reasoning on XML documents: A description logic approach. *JLC: Journal of Logic and Computation*, 9.

Colmerauer, A. (1984). Equations and inequations on finite and infinite trees. In *Proc. of the International Conference on Fifth Generation Computer Systems (FGCS-84)*, 85-99. Tokyo, Japan: ICOT.

Cohen, S., Kanza, Y., Kogan, Y., Nutt, W., Serebrenik, A. and Sagiv, Y. (1999). EquiX–easy querying in XML databases. In *Proceedings of the ACM SIGMOD Workshop on the Web and Databases (WebDB'99)*. Philadelphia, PA, June.

Decker, S., Brickley, D., Saarela, J. and Angele, J. (1998). A query and inference service for RDF. In *QL'98-The Query Languages Workshop* (http://www.w3.org/TandS/QL/QL98/). World Wide Web Consortium.

Deransart, P., Ed-Dbali, A. and Cervoni, L. (1996). *Prolog: The Standard.* Springer Verlag.

Grosof, B. N., Labrou, Y. and Chan, H. Y. (1999). A declarative approach to business rules in contracts: Courteous logic programs in XML. In *Proceedings 1st ACM Conference on Electronic Commerce (EC-99)*. Denver, Colorado.

Hanus, M. (1997). A Unified Computation Model for Functional and Logic Programming. In *Conference Record of POPL '97: The 24th ACM SIGPLAN-SIGACT Symposium on Principles of Programming Languages*, 80-93. Paris, France.

Harold, E. R. (1999). *XML Bible*. San Mateo, CA, USA: IDG Books Worldwide Inc.

Heflin, J., Hendler, J. and Luke, S. (1999). SHOE: A knowledge representation language for Internet applications. *Technical Report CS-TR-4078*, University of Maryland, College Park.

Krause, T. (1990). Klassifizierte relational/funktionale Klauseln: Eine deklarative Zwischensprache zur Generierung von Register-optimierten WAM-Instruktionen. *Technical Report SWP-90-04*, University of Kaiserslautern, Department of Computer Science.

Lloyd, J. W. (1987). *Foundations of Logic Programming*. New York: Springer Verlag.

Chapter IV

Four Different Types of Classification Models

Hans Rudolf Straub
Project Group for Semantic Analyses in Healthcare, Switzerland

INTRODUCTION: THE ROLE OF CLASSIFICATION MODELS

EDP stands for electronic data processing. Data bear imminent information–but what kind of information? The question becomes particularly apparent when a great deal of data is to be processed. In this case, the data have to be structured in order to give evidence of the information they bear. Thereby it is important: The primary data collected from a series of discrete observations are more comprehensive than any derived interpretation. They carry more information than any subsequent analysis or interpretation requires or can express.

Due to the reduction in the amount of data (and bits), the process of interpretation brings about a loss of information. It is only by structuring of the data that a proposition can be derived, and this brings about a loss of some primarily available information (Figure 1).

What kind of information is omitted? The question is not rhetorical at all, since it is the categorization which determines both the kind of information that is made explicit and the kind of information that is omitted. One might be tempted to anticipate a pre-existing, objective categorization. This is not the case, since the categorization is not only based on the data under consideration, i.e., the "object," but also on the aim of the study. The latter refers to the research question as posed by the investigator, i.e., the "subject." Depending on the research question, the point of view changes, resulting in a different way of extracting the relevant information.

Figure 1: Coding as an interpretation process

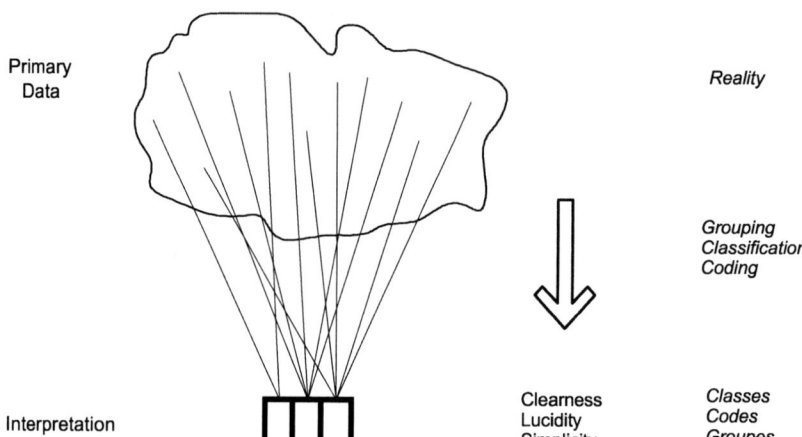

In order to preserve as many details and possible views as possible, while providing for a maximal clarity, the structuring of the data for interpretation and, thus, the applied model for knowledge representation are pivotal. In other words, *the product of clarity x granularity* must be optimized. This contribution is about the inner architecture of knowledge representations for classifications. It presents four different, successively refined models, the fourth of which allowing for a modeling of concepts which is close to the real world.

The value of classification models in the medical field cannot be overestimated. The amount of medical data gathered in hospitals is huge, and a systematic interpretation would substantially contribute to the creation of knowledge. The reason why such a systematic interpretation is not made on a routine basis becomes obvious: Information about treatments and diagnoses is provided as free texts and cannot be analyzed by mathematical (computational) methods without a proper classification of the underlying concepts. Because of the rich semantics of medical information, the classification model should not be too simple. It should be as detailed and close to reality as possible and, at the same time, clear and easily processable by computers. This chapter gives an overview of the strengths and weaknesses of the different classification models. The rising interest of public health in epidemiological data and of medical payment systems in fair Diagnosis Related Groups (DRGs) reflects the growing need to handle classification systems in a comprehensive way.

ONE-DIMENSIONAL, HIERARCHICAL MODEL

Confer with Table 1 for the characteristics of the four models.

Concepts can be regarded as "drawers" where similar objects or events of the real world can be stored (Figure 2). Thereby, the terms used to denote the concepts are the labels of the drawers. Of course a label can also be a number or a code instead of a name. Codes have the advantages of weaker memory requirements (which was more important with the first computers than today) and of encryption (which is important for military purposes). It might appear that codes have also the advantage of a systematic labeling. However, the systematic labeling results from the composition of the system of drawers and not from its attribution with labels.

The one-dimensional architecture lists the concepts one by one. Thereby, some kind of order is already created. In a further step, adjacent drawers can be merged to higher-ordered units on a superimposed level, i.e., the concepts are aggregated to superordinated concepts (Figure 3). The process can be repeated and the superordinated concepts can be aggregated as even higher-ordinated concepts. In so doing, a hierarchical tree is created from a linear sequence (Figure 4).

Thereby, the superordinated concept is implicitly provided by the system. It must not be explicitly stated for a particular concept. For instance, the code A00.1 (El-Tor-Cholera) of the ICD-10-Code implies:

- the superordinated concept to A00.1, i.e., A00 (Cholera)
- the superordinated concept to A00, i.e., A00-A09 (infectious disease of the intestine)
- the superordinated concept to A00-A09, i.e., A00-B99 (particular infectious or parasitic disease)

These implications do not have to be stated in a particular case. They are provided by the system along with the subordinated concept (El-Tor-Cholera) and cannot be omitted. The superordinated concepts contain no additional information than the subordinated concept.

Table 1: Survey of the four generations and their characteristics

Generation	Dimensions / Axes	Focus	Points
1.Generation	unidimensional=uniaxial=hierarchy	-	single-point
2.Generation	multidimensional=multiaxial	unifocal	single-point
3.Generation	multidimensional=multiaxial	multifocal	single-point
4.Generation	multidimensional=multiaxial	multifocal	multi-point

Figure 2: One-dimensional, hierarchical model

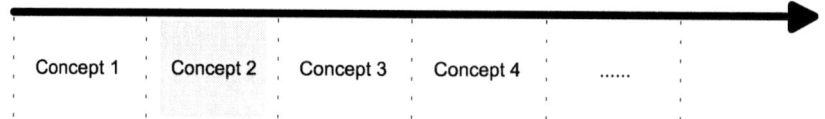

Figure 3: Aggregation of concepts to superordinated concepts

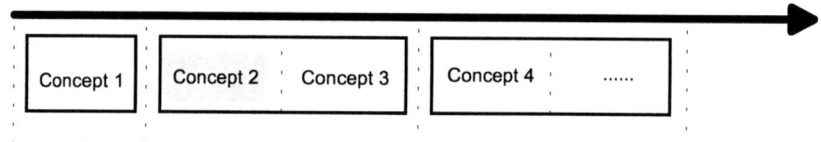

Figure 4: Creation of a hierarchy of concepts from a linear sequence

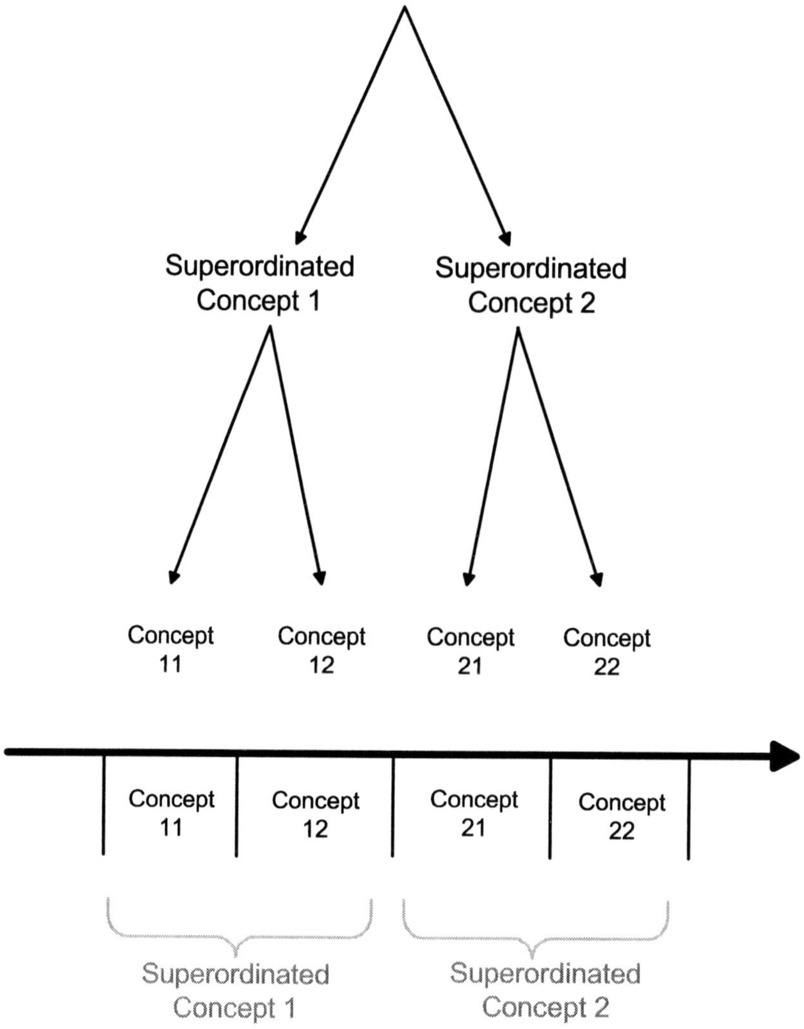

Superordinated
Concept 1

Superordinated
Concept 2

Concept
11

Concept
12

Concept
21

Concept
22

Concept
11

Concept
12

Concept
21

Concept
22

Superordinated
Concept 1

Superordinated
Concept 2

Thus, the hierarchical system presupposes a taxonomy which goes beyond the naming of the atomic concepts. In a particular case, e.g., in data transmission or storage, it suffices to state the last subordinated concept, i.e., the outmost leave carrying all superordinated concepts immanently. Although, at a first glance, the tree hierarchy appears to be two-dimensional, it can be easily mapped to a one-dimensional list. This mapping accounts for the easy handling of the hierarchical model by computer systems.

The ICD-10-Code is a nice example for the close relationship between the one-dimensional, linear code and the hierarchical classification. Indeed, most of the codes used in healthcare are of a hierarchical, one-dimensional type.

MULTIDIMENSIONAL, UNIFOCAL MODEL

The axis of the above described one-dimensional model, as depicted in Figure 2, can be regarded as a dimension in a space. The space of ICD-10 is the linearly sorted range of the "numbers" A00.0 to Z99.9. Each concept corresponds to a point in this space, on this axis, in this dimension. (Of course, "A" is a letter and not a number. However, from an information theoretical point of view, the only difference between letters and numbers is that the value range of the former consists of 26 values instead of 10. The linear, discrete, one-dimensional order is preserved with this system.)

Provided the system is unambiguous, any atomic concept is to be represented by exactly one value and any superordinated concept by an array of adjacent values. Any atomic concept must not be represented by two or more independent values. Distinguishable codes or values must be mutually exclusive, at least on the same axis (principle of *disjunctivity*). The requirement of disjunctivity enforces the use of unique codes. On the other hand, it has two disadvantages:

- The number of describable states is identical with the number of concepts (or codes). By using a single axis, it is not possible to get an overplus of states. In order to describe the real world approximately precise, an inflation in the number of drawers is inevitable, and these are no longer easy to handle.
- In order to represent a hierarchical system, an agreement on the semantic class at the top level is required. This semantic class defines then the top level of the concept hierarchy. No other semantic class can define superordinated concepts on that level. If two semantic classes overlap and, hence, can be combined (example below) either unnecessary repetitions would be generated or the disjunctivity principle weakened.

In order to illustrate the second disadvantage, a glance on the table of contents of the ICD-10-Code can be helpful. Chapter I comprises infectious diseases, Chapter II neoplasms, and Chapter IV metabolic disorders. These are pathophysiological groups which are more or less mutually exclusive. In contrast, Chapters III, VI, VII, VIII, etc., refer to particular organs or tracts (which are likewise mutually exclusive) such as eyes, ears, cardiovascular system, etc. Thus, we have to deal with two types of mutually exclusive top-level concepts. (If the analysis is extended to all chapters, of course, more than two types can be identified. However this does not weaken the presented implications.) These types can be combined. Both infectious diseases of the ears and neoplasms of the eyes occur. But where in the tree hierarchy can they be found? Among the neoplasm's or the eyes' diseases? This lack of clarity on the top-level affects the lower levels and is one reason for the many inconsistencies the user of such a pseudo-hierarchical code faces. ICD-10, which has a long tradition and grew over time, is quite conscious about this lack of clarity. It tries to guide the user by a number of "exlusiva," "inclusiva" (i.e., by extensional enumeration; Ingenerf, 1993) and further commentary. These efforts are crowned with the attempt to outsmart its own hierarchical system by the "cross-asterisk" classification (DIMDI, 1994/95, Volume II). Whether the handling is facilitated by the complex body of legislation and the very inconsistently implemented cross-asterisk approach must be questioned.

The second type of classification model anticipates the first disadvantage (with a number z of concepts not more than z different states can be described) and, in a way, also the second disadvantage (problem of commingling types). It introduces the property of multidimensionality (Figure 5).

In Figure 5, the universe of concepts is classified along two semantic classes (A and B). Similar to two dimensions, the semantic classes span a plane. Each place on the plane (i.e., each field) is associated with any of the two dimensions in such a way that it is assigned a precise value in each dimension. Conversely, for every combination of dimension values there is exactly one field. This setting has several advantages over the linear model. As an example, imagine the semantic class A to refer to infectious agents and the semantic class B to organs of the body. Possible values for the semantic class A could be:
- staphylococci, tuberculosis bacteriae, pneumococci, influenza viruses, HI-viruses, meningococci, gonococci, plasmodiae malariae, etc.

For the semantic class B the following values would be possible:
- Blood, lung, meninges, bone, skin, kidney, liver, etc.

Each field which is described by these two semantic classes combines a value A with a value B, e.g.,

Figure 5: Multidimensional, unifocal model

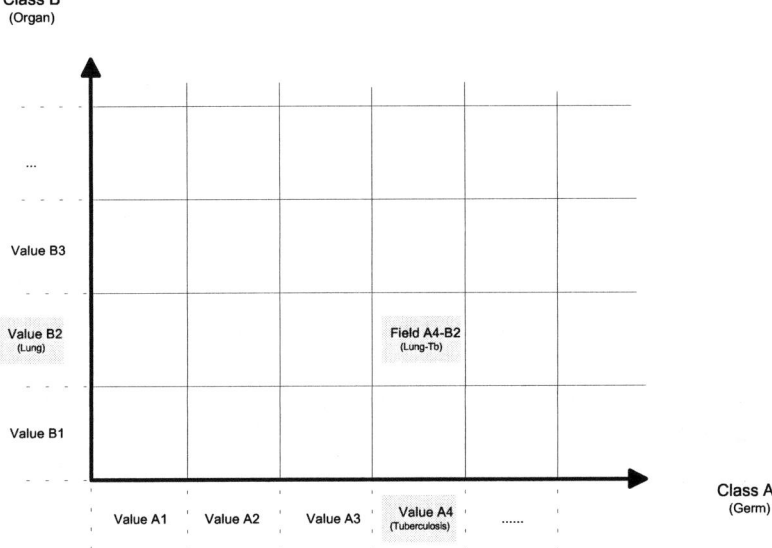

- staphylococci / lung -> staphylococcus pneumonia
- tuberculosis / bone -> bone tuberculosis

In the multidimensional, unifocal model, two semantic classes A and B can have two independent values (Wingert, 1984, denotes independent semantic classes as orthogonal). The two semantic classes (or *axes* or *dimensions* or *degrees of freedom*; the four terms are almost used as synonyms in this contribution) span a two-dimensional plane. The number of fields corresponds to the number of states of the real world which can be described by the model. Provided that a is the number of values of semantic class A and b the number of values of semantic class B, then the model can describe, with $a + b$ different values or concepts, $a * b$ different states. In other words: There is a propagation in the number of describable states compared to the number of concepts required for the description. This propagation corresponds to the shift from addition to multiplication. With the 16 concepts listed in the example (8 of semantic class A and 8 of semantic class B), 64 (8 * 8) real-world-related states (fields, combinations of semantic classes) can be described. By the use of more axes and more values per axis the propagation becomes even more apparent. A uniaxial concept model, such as ICD, would list each combined state one by one, thereby getting difficult to handle and glutted.

Basically, the number of axes is not limited. A three-dimensional model can describe, with $a + b + c$ different concepts, $a * b * c$ different states. By the use of four semantic classes the number of fields is even higher. The more dimensions are introduced, the richer are the description options.

One of the most important questions concerning the multidimensional model is which semantic classes should be selected as dimensional axes. SNOMED (Standardized NOmenclature of MEDicine), for instance, is a very elaborate multidimensional code system, originally of the pathologists. Its second version comprises seven axes (the current version 3 comprises 12 axes, concerning the inflation in the number of axes see the next section):

- M-code: morphology: what kind of tissue damage is observed (inflammation, neoplasm, etc.)
- E-code: etiology: what is the cause (e.g., infectious agents)
- T-code: topology: what is the localization (e.g., lung)
- F-code: function (e.g., hyperventilation)
- D-code: disease (e.g., shingles)
- P-code: procedure (e.g., resection)
- J-code: job (e.g., goldsmith)

The variability in the descriptions of the real world by the multidimensional code has the slight disadvantage of describing a given state not by a single code but by a combination of codes. Whereas in the case of ICD a single code suffices to describe a disease, with SNOMED one, two, or more are necessary (note that in order to describe a diagnosis, not all axes must contribute a value).

Difference Between the Hierarchical and the Multidimensional Systems

The denotation of three codes or concepts in SNOMED does not signify the same as the denotation of a concept and its superordinated concepts in a hierarchical system. In both cases, a number of concepts are denoted. In the hierarchical system the additional concepts are superordinated concepts, which can be implicitly derived from the subordinated concept. In contrast, in the case of a SNOMED triplet the three concepts are independent.

A one-dimensional system can appear to be multidimensional particularly, a hierarchy can look like a two-dimensional system. However, this two-dimensionality is only seeming, since there is no additional information contained in the additional dimension (see above). In contrast, a real multidimensional system contains independent information in any dimension.

In order to illustrate this: The axes of a real two-dimensional system can be considered as one-dimensional subsystems with their own hierarchies. Figure 6 shows a two-dimensional system with the two semantic classes organ and infectious agent.

MULTIDIMENSIONAL, MULTIFOCAL, SINGLE-POINT MODEL

The above described simple multidimensional model has two disadvantages:

- "Irrelevant" fields, i.e., fields which cannot be filled (Wingert, 1984) can occur.
- It is not easy to agree on a definite number of axes.

For the axes the same holds as described above for the concepts (or drawers or fields). If there are too many, the system gets difficult to handle; if there are too few less, it gets inaccurate.

Figure 6: Two-dimensional system and one-dimensional subsystems

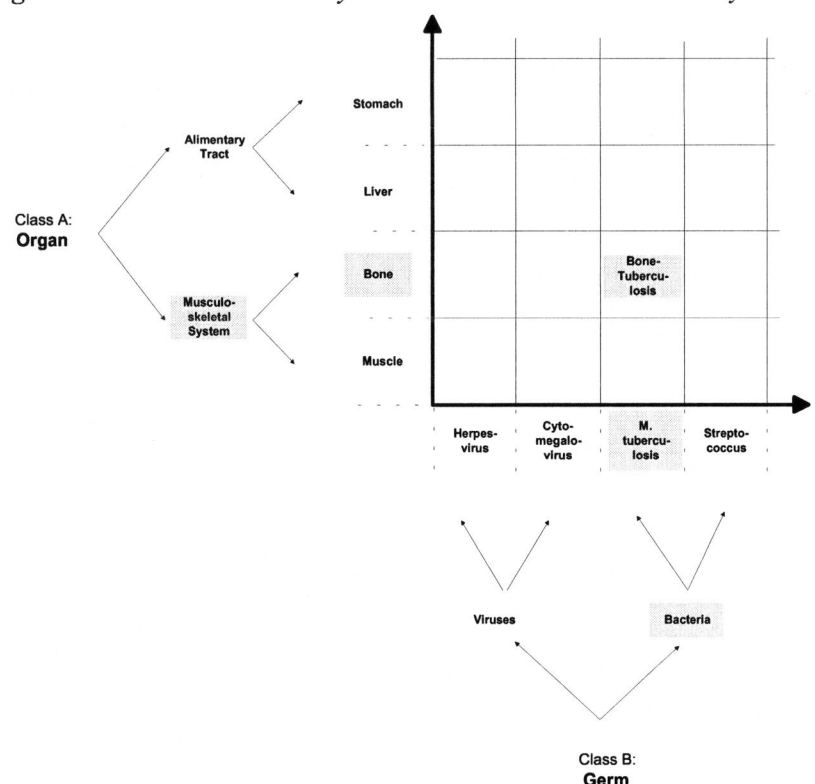

Inaccuracy of an axis means that it is not really "linear." That is, the values (concepts) on the axis are not of the same type. This results in an ambiguity of the superordinated concepts constructed by the hierarchy of the axis. What is happening is exactly the same as described in the preceding section as a disadvantage of the one-dimensional model. There, it was suggested to untangle the types and to increase the number of axes. Thus, in the case of inaccuracy the number of axes must be increased. The questions are whether or not the number of axes will grow to infinite and how a huge number of axes can be handled.

In the two-axial system, concepts can be allocated to sections on the two axes or to fields (Figure 7). In the three-dimensional system they can also be allocated to cubes. A 12-dimensional cube and 12-dimensional concepts can hardly be imagined, all the more since the latter would have twelve 11-dimensional superordinated concepts each and each of these again eleven 10-dimensional superordinated concepts, etc. Nevertheless, such highly dimensional constructions are inevitable, not only from a theoretical point of view, but also from the point of view of the practice of the medical terminologies. The complex medical terminology is even the normal case. Fortunately, the described problem with the large number of dimensions has a quite easy solution. Thereby both disadvantages, i.e., the irrelevant fields and the difficulties with the handling of multiaxial models, are anticipated.

Irrelevant fields occur if combinations of values do not make sense. For instance, *Fracture* is a value on the axis *Diagnoses* (Figure 8). This value can be combined with one out of two values (*Open* and *Closed*) on a second axis denoting the state of the skin barrier. (If the skin barrier is open, dirt particles and infectious agents can penetrate and cause inflammations. Therefore, fractures with open skin barrier reconstitute worse. Hence, the semantic class *State of Skin Barrier* is important.) *Diabetes* is a further diagnosis, but the values *open* and *closed* are hardly combinable with it. That is, it makes no sense to talk about a

Figure 7: Three different multiaxial systems

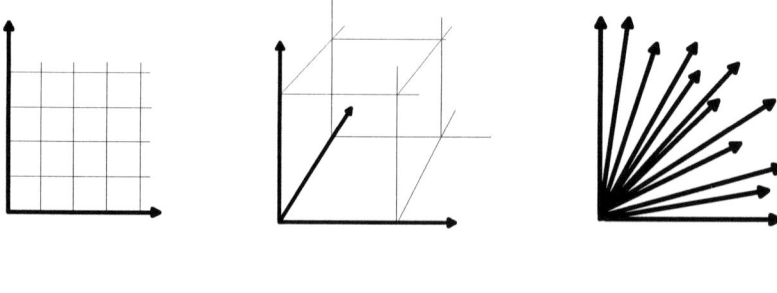

2-Dimensional System   3-Dimensional System   12-Dimensional System

Diabetes with Open (or Closed) Skin Barrier.

In Figure 8, the points D.o. and D.c. *(Diabetes with Open (and Closed) Skin Barrier)* are senseless (as opposed to the points F.o. and F.c.) but the multiaxial representation generates them. A possible solution is shown in Figure 9.

In Figure 9, the values *Open* and *Closed* are integrated in the axis with the diagnoses resulting in the generation of the superordinated concept *Fracture* and of the subordinated concepts *Open Fracture* and *Closed Fracture.* However, this solution is extremely shortsighted.

Fractures have semantic classes in addition to the state of the skin barrier, such as the involvement of joints, the number of bone fragments, the affected bone, etc. These semantic classes can be arbitrarily combined and, therefore, cannot be represented by a *single* hierarchy. They are, in a true sense,

Figure 8: Occurrence of irrelevant fields

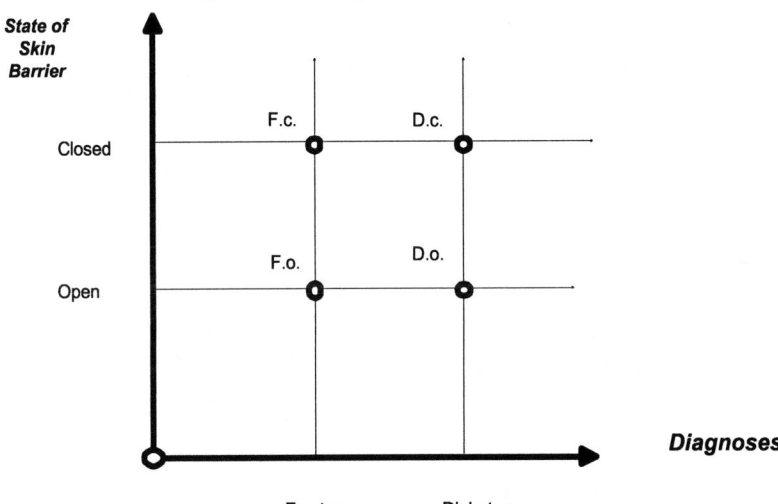

Figure 9: Intuitive solution to the problem of irrelevant fields

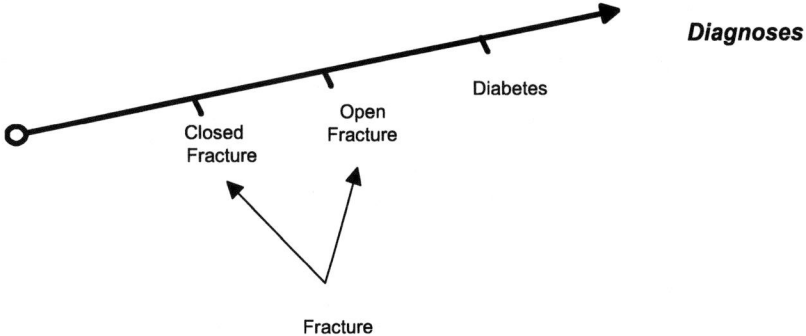

multidimensional. (The ICD-10 simulates for the semantic class *Open/Closed* the multidimensionality by an optional fifth position. However, because of its basic hierarchic structure it cannot provide all required semantic classes similarly with an additional position.)

As a consequence, the reduction to a single hierarchy, as shown in Figure 9, must be abandoned. The solution lies in maintaining the additional axis, but in a model, in which not all axes intersect in one *central* point. Instead, the axis is placed on a particular value (or on several values, i.e., the same axis can be added to several foci), thereby creating an additional *focus*.

In Figure 10, the axis *Skin Barrier* (with the values *Open* and *Closed*) is placed on the value *Fracture* of the axis with the diagnoses. The value *Fracture* is, in a sense, the focus for this (and further) axis (Figure 11). In contrast to the simple multidimensional model, this kind of representation not only allows for a single, central focus, but also for concurrent foci which are intersection points of axes. (In Figures 10 and 11 the focus is represented as a little circle. The stronger circle on the left represents the central focus, i.e., the center of the multifocal system.)

By shifting the axes from the center to the periphery, the disadvantage of the difficult handling of a multiaxial system is anticipated. Each axis is placed where it must be.

The second disadvantage, i.e., the occurrence of irrelevant fields *(Diabetes with Open Skin Barrier)* is anticipated by restricting the possible combinations of axes to those combinations for which the semantic classes are explicitly foreseen. In the same way as the requirement of disjunctivity can be anticipated by allowing for multiple axes, the requirement of (potentially) exhaustively filled fields can be anticipated by allowing for multiple foci.

Figure 10: Additional axis assigned to a focus

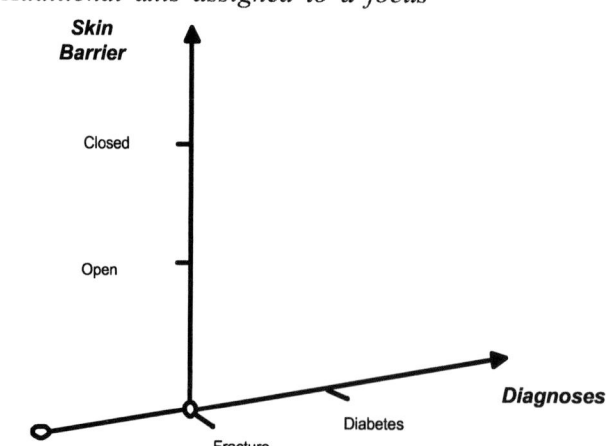

Figure 11: Three axes assigned to a focus

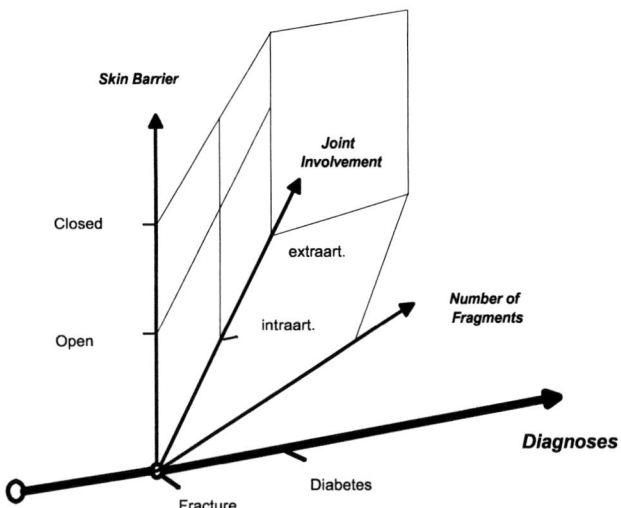

Disjunctivity and exhaustively filled fields are opposite requirements. It is only by anticipating both that a good spatial representation of the concepts is achieved. The concepts and their relations are represented in a model and allow for a spatial modeling of a domain which is close to the real world.

Applications of the Multiaxial, Multifocal Model

Systems of the described third generation are very common (however, they are not common among medical classifications):

- relational databases (RDB)
- object-oriented programming (OOP)
- frame-systems

Sowa (1984, 2000) demonstrates the relationship among the internal structures of the concept representations in these three systems. Of course, RDB, OOP and frames are three different "worlds" with different intentions, strengths, and opportunities. Nevertheless, as far as the structure of their data concepts is concerned, they are related and can be mapped to each other. The data concepts are the *infostatics*, i.e., the "system of drawers," the classification model, which is the focus of this chapter. The handling of methods, which separates OOP from the others, is not an issue of infostatics, but of *infodynamics*. Infodynamics denotes the handling of data over time, i.e., the communication, processing, and modification of data. The way in which the data are structured at any time point is infostatics, and herein RDB, KL-1, and OOP are not different (the same applies to inheritance, which is likewise a feature of the third-generation model and of infostatics; a detailed discussion of inheritance would exceed the scope of this chapter). However, the three are equally different from both the simple mono-hierarchical and the multidimensional, uni-focal models. RDB, KL-1, and OOP are examples for the third generation of data structures they are multiaxial, multifocal systems.

In consideration of the extension and complexity of the medical terminology and of the significance of healthcare for our lives, it is astonishing that to date few attempts have been made to relate the medical classifications and terminologies closer to the real world as, e.g., ICD-9 and ICD-10 do. There are many reasons why this is not the case. At a first glance, a mono-hierarchical system seems obvious. SNOMED, MESH (MEdical Subject Headings) and UMLS (Unified Medical Language System) are attempts to overcome such simple mono-hierarchical concept models. However, even the hereby shown classifications of the third generation are still not perfect. Medical concepts cannot be fully represented by models of the third generation. The reasons therefore and the method of resolution are presented in the following section.

MULTIFOCAL, MULTI-POINT MODEL

As shown, the requirement of disjunctivity motivated the shift from the mono-hierarchical to the multidimensional model. On the other hand, the requirement of exhaustively filled fields limited the scope of the dimensions and allocated the axes to a capsulated (focal) scope. When working with these systems, there arise still problems in connection with the requirement of disjunctivity, which cannot be handled by the introduction of additional axes. The following example will illustrate this.

Bone fractures are associated with particular bones. Thus, the fractures of the forearm can be divided into those of the ulna and those of the radius. Furthermore, fractures of both bones can be the result of one and the same accident. In this case, there is one hospitalization, one operation, one invoice, and one diagnosis: double-fracture of the forearm (Figure 12).

Of course, the double-fracture of the forearm is not the same as the sum of the fractures of radius and ulna. The patient gets only one (instead of two) hospitalization, one operation, and one anesthesia. Regarding the costs thereby incurred as well as the medical implications, this differentiation is much more than a sophistic bauble. It becomes clear: The whole is not identical to the sum of its parts. Therefore, the double-fracture of the forearm can be assigned a distinct value on the axis *Diagnoses*.

Although it is now possible to separate double-fractures from isolated

Figure 12: Fractures of the forearm

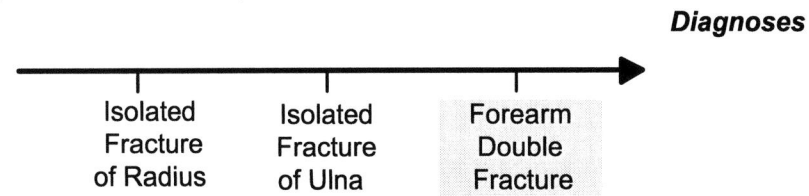

fractures of the ulna, this classification is still not satisfying. The principle of disjunctivity appears to be anticipated, and all fractures can be unambiguously assigned. However, if for any reason, after a successful assignment, all fractures of the ulna must be retrieved from a record of diagnoses, the double-fractures must be included as a special case (Figure 13).

Thus, with the data analysis the disjunctivity problem slips in through the back door. But this is not all. The double-fractures are not only a special case of the fractures of the ulna but also of the radius. Figure 14 shows a Venn diagram of the valid combinations.

In Figure 14, the two sets of fractures of ulna and radius intersect, and the set of double-fractures of the forearm corresponds to the overlapping area. Such is the real-world case. However, thereby the requirement of disjunctivity is not anticipated. By means of the above Venn diagram, be it reflected in the human mind or implemented as a value triple on the diagnosis axis in a computer system, it is possible to represent the real world properly. However,

- The principle of disjunctivity, which requires that the objects assigned to values on an axis of a given semantic class form independent sets, is violated. The requirement of disjunctivity is not only a theoretical postulate. Not until an unambiguous assignment can be achieved in practice a classification makes sense.

- The dependencies among the sets would have to be explicitly defined. Everyone who classifies a given case must be aware of the rules; if not, he classifies wrongly. For instance, he/she must not classify an isolated fracture of the ulna if also the radius is broken and vice versa. In both cases, he/she must assign the fracture to the double-fracture. In addition, also the one who makes the analysis must be aware of the rules. If he/she is looking for fractures of the radius, he/she must include the double-fractures of the forearm and think about further cases he/she might have forgotten.

- Multiple fractures and traumata are not at all rare, statistically negligible exceptions. In contrast, they are very frequent. A lot of them have proper names (e.g., Monteggia fracture, unhappy triad, Weber fracture). Along with infractions of bones, often also the skin,

Figure 13: Fractures with involvement of the ulna

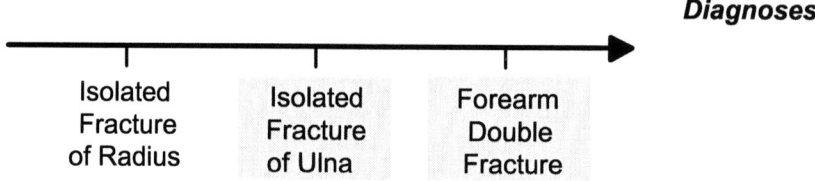

Figure 14: Venn diagram of forearm fractures

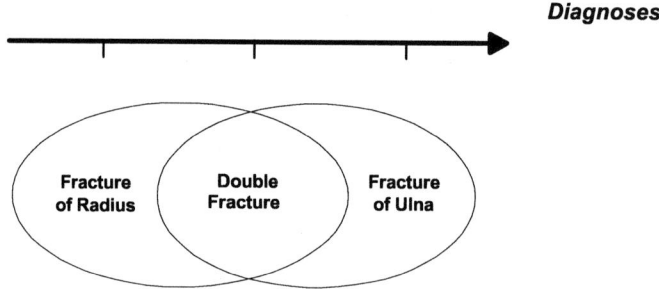

nerves, and tendons are injured. In the domain of internal medicine it is all the same. Multi-morbidity is the rule. A lot of diagnoses occur almost regularly in combination (e.g., adiposity, diabetes mellitus, hypertension, and consequential damages constitute the so-called metabolic syndrome). A lot of diseases are only known as syndromes, i.e., combinations of symptoms.

• Since multi-morbidity and multiple fractures are the rule, the proper classification of diagnoses is very difficult and errors on the part of the coding person are almost inevitable. The resources required for this task are heavily underestimated.

The representation of Figure 14 is not practical in a further respect: The localization is not the only semantic class of a fracture. Fractures have a lot of further semantic classes (such as the condition of the skin barrier), which can be combined with the localization in almost any way. As mentioned in the presentation of the multifocal type, it makes sense to consider *Fracture* as a focus, in which all axes relevant for that focus intersect, among others the axis which denotes the broken bone. This results in Figure 15.

However, something is wrong: On the *Diagnoses* axis it was possible to imagine a value for the double-fracture. In contrast, on the *Bone* axis a value for *Double-bones* makes no sense. In other words, the principle of disjunctivity must be weakened, but in a controlled and selective manner (Figure 16).

As opposed to the three models so far discussed, the multi-point model allows a semantic class (an axis) to contribute at the same time more than one value. Thereby, the concepts must not be adjacent, i.e., they must not be subordinated concepts of the same superordinated concept. Sets of arbitrary values from the same axis can be bundled and form in their entirety the value of this semantic class for a substantial entity.

This weakening of the disjunctivity and unambiguousness within a semantic class must be recorded in the system. Therefore, in Figure 16, the repeated naming on the *Bone* axis is indicated by the value *2* on the axis *Number of Bones*. The system must know the most common multiple diagnoses. These must be available as "aliases." Primarily the persons performing the statistical analysis must know that there are different levels of understanding. This knowledge can be conveyed in different ways, similar to the different specializations of the above discussed multifocal model, i.e., KL-1, OOP languages, RDBs, particle model of the CSL (Straub, 1994). Without appropriate consideration of the multiple points, no classification system can persist.

Figure 15: Forearm fractures in the multifocal (single-point) model

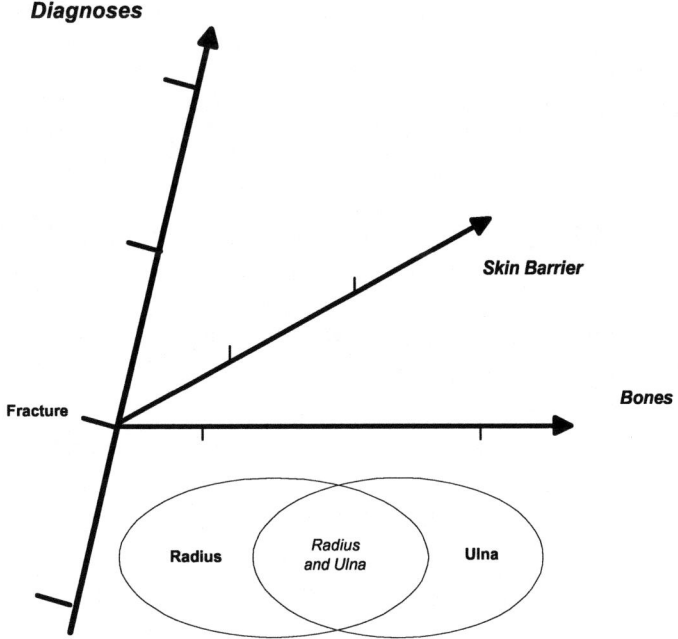

Figure 16: Multifocal, multi-point model

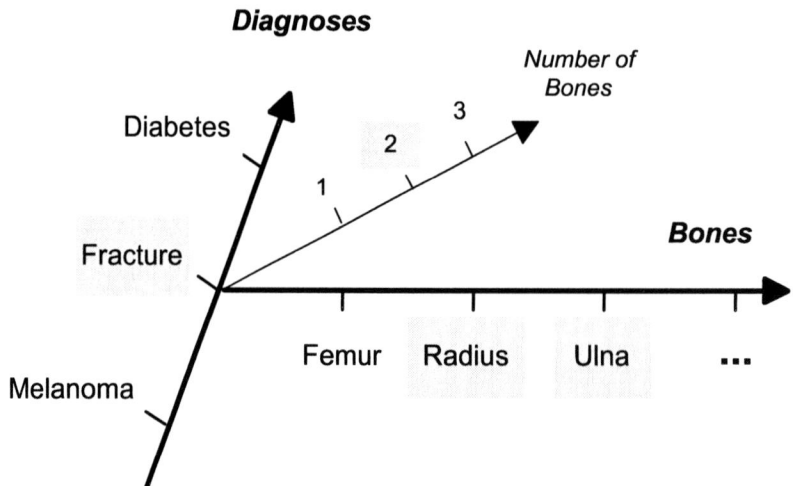

Difference Between the Multiaxial and the Multi-Point Solutions

Why do multiple points have to be introduced? The requirement of disjunctivity already led to the introduction of a new model. Why can the thereby proposed solution, i.e., the introduction of additional axes, not be applied to the current problem?

In order to answer these questions, it might be helpful to consider the relationships among the sets and the representation of the sets in a spatial framework. Let us first consider the example which led to the introduction of multidimensionality (Figure 17).

Both meningitis and tuberculosis are sound medical diagnoses. However, they overlap, i.e., they are not disjunct. The intersection is the tuberculous meningitis (Figure 18).

If, as shown in Figure 18, the sets are allocated to two axes, an equivalence of the relationships among the sets and the axes is largely achieved. In other words, it is possible to properly map the existing relationships among the sets on a coordinate system by introducing an additional axis.

In the case of the forearm fractures the things are fundamentally different (Figure 19). On a first glance, there is an apparent similarity to the previous case: Fractures of the ulna and of the radius are both sound medical diagnoses. Both belong to the same class (i.e., the fractures). They are not completely disjunct and their intersection is the double-fractures of the forearm. So far the analogy holds. In the current case, the *disjunctivity is of a fundamentally different nature.* Particularly, the disjunctivity *cannot* be anticipated by an increase in the number

Figure 17: Meningitis and tuberculosis in a one-dimensional layout

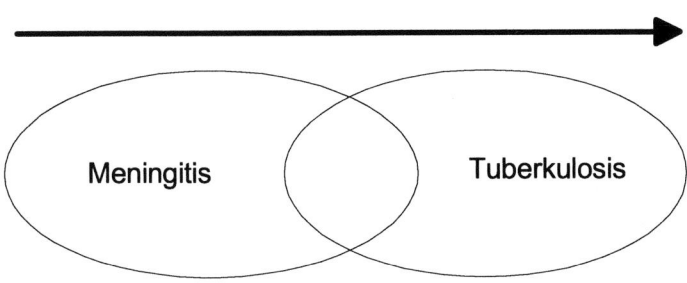

Figure 18: Meningitis and tuberculosis in a two-dimensional layout

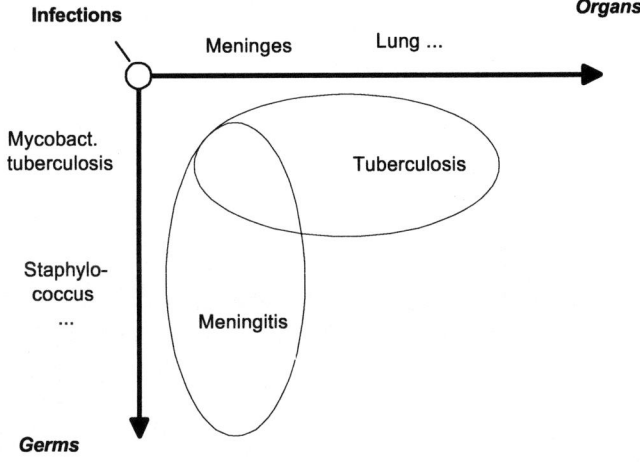

Figure 19: Fractures of ulna and radius in a one-dimensional layout

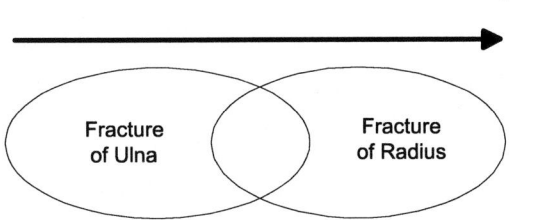

of axes.

Figure 20 shows the abortive attempt to map the relationship among the sets in a two-axial coordinate system.

The horizontal axis must allow for the distinction between the intersection (double-fractures of the forearm) and the remaining fractures of the radius, similar to the distinction between tuberculous meningitis and the remaining tuberculoses in Figure 18. There the semantic class *Organs* was considered, whereas, in this example, the distinction considers the number of broken bones. Similarly, the distinction on the vertical axis considers the number of broken bones of the fractures of the ulna.

However, the labeling of the axes, as proposed in Figure 20, makes no sense. This is shown by the cross which denotes a logically inconsistent field. More precisely, it denotes a fracture of a single bone which, at the same time, is a fracture of both the radius and the ulna. In contrast, in Figure 18, the corresponding field is free from any logical inconsistency. Moreover, it denotes a real diagnosis, i.e., the staphylococci pneumonia.

There are further logical inconsistencies. The upper quadrant on the right denotes an isolated fracture of the radius. According to the two axes, it would also denote a fracture of the ulna with two broken bones! And where would a fracture of the upper arm have to be represented? In contrast, in Figure 18, such inconsistencies do not appear. Although there are combinations which

Figure 20: Attempt to map fractures of ulna and radius in a two-dimensional layout

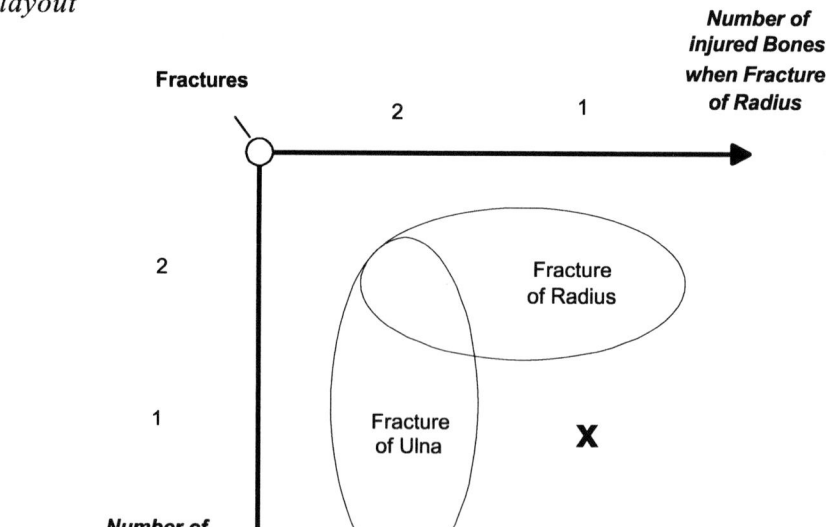

do not really occur (e.g., influenza viruses in the skin), they are not completely impossible and, above all, there are no logical inconsistencies in the coordinate system similar to those in Figure 20.

One might object that, in Figure 20, the axes are poorly defined. However, the reader is invited to define alternate axes which do not lead to inconsistencies. The described observations support the assumption that there are fundamental variations in the nature of disjunctivity. Whereas, in a first class of cases, the requirement of disjunctivity can be anticipated by introducing additional axes, it can, in a second class, only be anticipated by multiple points as presented in this section.

THE ROLE OF CLASSIFICATION MODELS (REPEAT)

Why do we make classifications? If we have to deal with a huge number of heterogeneous objects in the real world, we must group them in order to treat similar cases the same way. Thereby, the complexity of the real world should be represented as precisely and simply as possible. The models of the fourth generation approximate the real world best (Figure 21).

If precision is not required, simpler models still have some advantages. They are intuitively comprehensible and easy to handle. Whenever decisions have to be made, a simple view prevents being confused by complexity. Likewise, not all aspects are equally important. Thus, only the most significant aspect is considered in order to classify the objects and unimportant details are omitted (Figure 22).

Simple classifications are certainly attractive, but their basic problem is shown in Figure 23.

Since the hierarchical, unidimensional classification is too much of a simplification, there are *always* alternate unidimensional classifications possible. This is the reason why it is so difficult to agree on a *single* (unidimensional) code. This is also the reason for the difficulties with the agreement on a common set of DRGs (Fischer, 1997). And if all parties have come to an agreement, it is often only seemingly, and modifications are soon made. Although the resulting classifications are simple, they are heterogeneous and jeopardize the primary objective of any classification, i.e., the comparison of data from different sources.

Nevertheless, a simple, even an unidimensional, classification can be reasonable, but only in view of a particular task, a particular objective (Straub, 2001; Straub & Mosimann, 1999). A different question results in a different view and in

Figure 21: Derivation of more simple models from the complex model

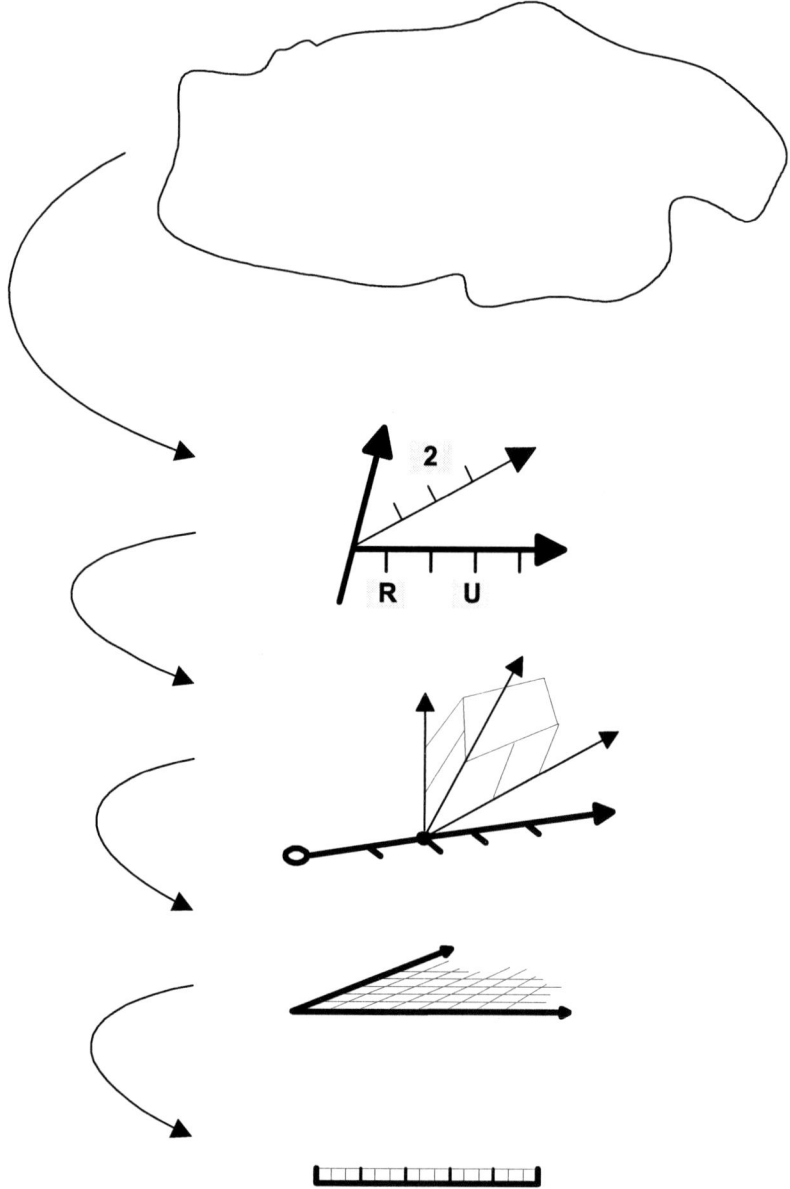

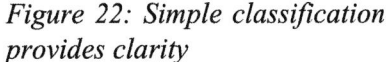

Figure 22: Simple classification provides clarity

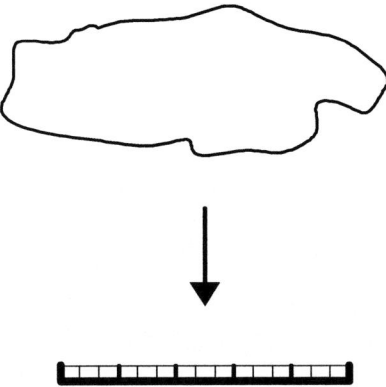

Figure 23: Alternate classifications confuse

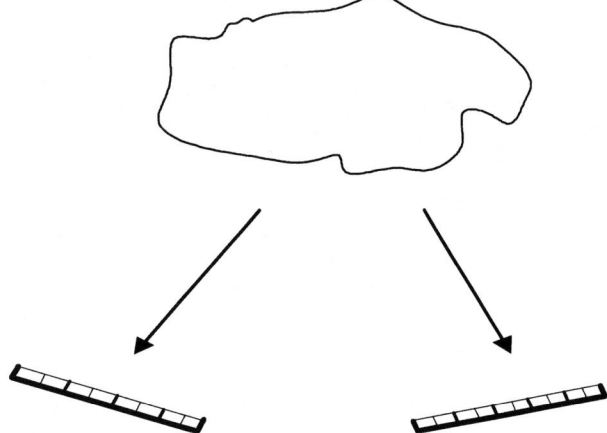

a different uni-dimensional classification.

Both the requirement of as precise a representation as possible, which is close to the complexity of the real world, and the requirement of as simple a representation as possible can equally be considered if classifications of different generations are mixed. Thereby, it is important that the different systems are applied at the right place (Figure 24).

In a first step, the real world is represented by the model, which is closest to it. Depending on the objective of the analysis, the collected information is then provided by simple multi- or uni-hierarchical classifications. If the information is initially entered in a consistent multifocal, multi-point classification system, subsequent mappings on arbitrary systems on a lower level can be easily achieved, since clear rules for these mappings can always be derived (of course, the opposite is not the case). If a complex classification is made at first, a subsequent interpretation is not restricted, since a flexible arrangement of the information according to the criteria of the particular task is always possible.

In sum, the presented allocation of concepts to classifications experiences increasing degrees of complexity and closeness to the real world (Table 2). Each of the systems of the first generation, i.e., the hierarchical systems, has a single degree of freedom. The multidimensional systems of the second generation have n degrees of freedom, one for each dimension. The degrees of freedom of the third generation are no longer independent from each other; they form a network.

Figure 24: Mixing architectures of different generations

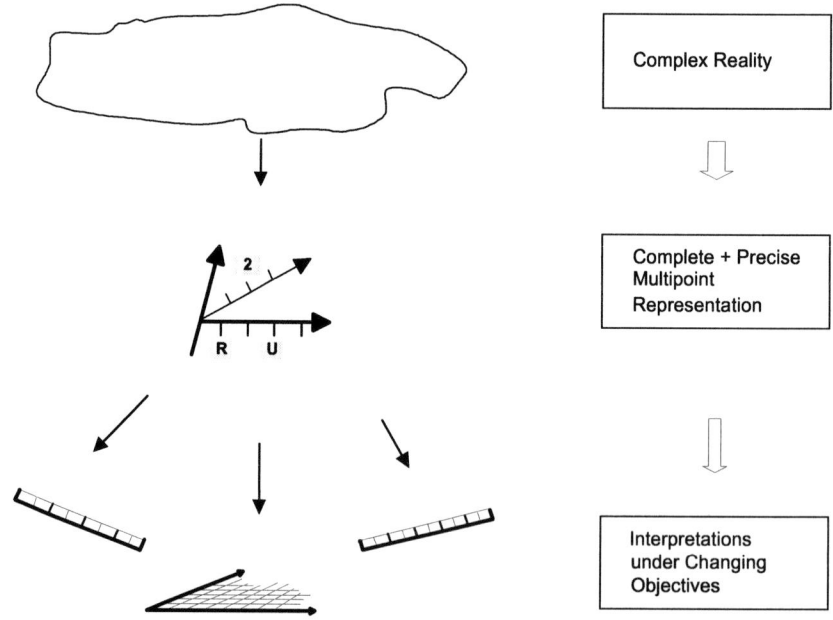

Table 2: Differences in number of axes, foci, and points between the four generations

Generation	Number of Semantic Axes	Number of Foci	Number of Points per Axis
1.Generation	1	0	1
2.Generation	n	1	1
3.Generation	n	z	1
4.Generation	n	z	m

Depending on the respective value on an axis, additional degrees of freedom (axes) are provided. Thus, the axes intersect in a number of points (foci), thereby accounting for the labeling of the systems of the third generation as multifocal. Considering the fourth generation, the web of axes (the semantic space) has the same structure as the third generation. Different from the latter, the fourth generation allows for an activation of concomitant values on a single axis, i.e., several values of the same type can be selected at the same time. By this means, composed objects, such as multiple diagnoses, can be represented in a logically consistent way, thereby allowing for a processing by machines.

REFERENCES

DIMDI. (1994/95). *ICD-10: Internationale statistische Klassifikation der Krankheiten und verwandter Gesundheitsprobleme* (10 Rev.). Bände I/II/III. München: Urban & Schwarzenberg.

Fischer, W. (1997). *Patientenklassifikationssysteme zur Bildung von Behandlungsfallgruppen im stationären Bereich.* Bern: Bundesamt für Sozialversicherungen.

Ingenerf, J. (1993). *Benutzeranpassbare Semantische Sprachanalyse und Begriffsrepräsentation für die Medizinische Dokumentation.* St. Augustin: Infix.

Sowa, J. F. (1984). *Conceptual Structures: Information Processing in Mind and Machine.* Reading: Addison-Wesley.

Sowa, J. F. (2000). *Knowledge Representation.* Pacific Grove: Brooks/Cole.

Straub, H. R. (1994). Wissensbasierte Interpretation, Kontrolle und Auswertung elektronischer Patientendossiers. *Kongressband der IX.*

Jahrestagung der SGMI, 81-87. Nottwil: Schweizerische Gesellschaft für Medizininformatik.

Straub, H. R. (2001). Das interpretierende System: Wortverständnis und Begriffsrepräsentation in Mensch und Maschine, mit einem Beispiel zur Diagnosecodierung. Wolfertswil: Z/I/M-Verlag.

Straub, H. R. and Mosimann, H. (1999). Codierung als interpretationsvorgang. *Schweizerische Medizinische Wochenschrift*, 129(Suppl 105/II), 275.

Wingert, F. (1984). *SNOMED Manual*. Berlin, Springer.

Chapter V

Towards Improved Representation and Communication of Pharmaceutical Expert Information[1]

Markus Wagner
University of Mainz, Germany

MOTIVATION

The main concern of healthcare management has ever been embodied by two conflicting goals. These are improving care quality while reducing care costs. Information technology already has proven its positive impact on both goals. While general administrative tasks are being performed more efficiently for some decades, expert-oriented systems are entering the medical domain with increasing support capabilities for domain-specific therapeutic procedures, leading to more secure and efficient therapy planning, monitoring and control.

The demands on quality and efficiency of care are posed by social, legal and political areas, resulting in growing expectation, uncertainty and pressure concerning healthcare professionals' use of expert information. These circumstances are cornering medical practitioners. A variety of role-specific information problems may be identified which finally can be attributed to the nonavailability of suitably represented expert information as well as to missing automatization of medical procedures. In their entirety they indicate two major information problems in healthcare.

The *representation problem* consists of the lack of a universal representation of pharmaceuticals which formally describes their common properties as well as their behavior within groups of simultaneously applied drugs. Internal representation affects granularity and precision of information as well as the functionality of the application system. External representation affects suitability of access and navigation mechanisms, which depend on the user's role and situation. An internal representation may meet the needs of many applications, while an external representation can only be as suitable as the internal representation allows.

The *communication problem* refers to the communication capabilities of a representation delivered by a solution of the previous problem. There is nearly no benefit of a pharmaceutical information model if information is not suitable for transportation. This means that the impact of an internal representation reaches interorganizational cooperation capabilities. This affects particularly the diversity of semantic reference systems used by local representations. The communication problem consists of finding means to enable organizations to exchange information with a minimum of human intervention.

As the information jungle continues to grow, healthcare costs and treatment quality change to worse. The patient has to bear the consequences. While his contribution on treatment costs increases, no one can guarantee that the medication designed by the physician is optimal in both therapeutic and economic senses. For instance, many expensive commercial-brand products of major manufacturers may be substituted by more beneficial, therapeutically equivalent generics, but without this information at his fingertips the physician will continue to prescribe products with concise and habitual names.

APPLICATION DOMAIN

The application domain targeted by pharmaceutical information systems includes any medical subject area affected by information on properties of pharmaceutical products. This comprises the treatment situation at the physician's desktop, drug dispensing procedures in hospital pharmacies, pharmaceutical consultation at the drugstore as well as patient home information. Although differences exist among role- and situation-specific external representations of user-oriented information, any application of pharmaceutical information systems relates to some properties of pharmaceuticals.

This universal applicability of pharmaceutical information systems leads to their architectural role as a subsystem within a not further specified superordinated medical information system. Many applications may exist (Hempel, Gräfe, Helmecke & Schuchmann, 1994). Thus, these systems are highly polymorphic in nature and the need for a single, universal internal representation can be justified with reusability, integrity and universality. The role leaves open any assumptions of the application domain. Any usage is specific to the superordinated application system. However, three principal applications of pharmaceutical information systems may be considered.

Medication Analysis is a procedure that aims to identify any critical problems within a given input medication. A set of problems is computed, each of which may be a contraindication, an interaction or an overdose. It is desirable that this procedure also adds links to further information found in local or external documents. Medication analysis can significantly improve security of drug dispensation (Zagermann, 1996).

Medication Transformation is a procedure that aims to construct a new medication which is therapeutically equivalent to an input medication, and which meets certain optimization criteria. The most beneficial criteria refer to cost reduction. Then, some of the prescribed drugs are substituted with the most beneficial but therapeutically equivalent products available. This procedure relates to managed care (prescription benefit management), as is practiced in the United States as an intervention of insurance companies in the prescriptions of physicians. Medication transformation can significantly improve prescription efficiency.

Medication Synthesis is a procedure that aims to construct a medication based on its desired properties. The input is a specification of these properties in terms of wanted and optionally unwanted effects, and the output is a medication with a maximum of conformity with the specification and a minimum of undesired properties, concerning both therapeutic and economic issues. Until now, medication synthesis is to be regarded as a theoretical application and is mentioned because it formally is the opposite procedure to medication analysis.

The application goal for the physician seems to be clear: The goal is to be able to define an optimal medication for a patient with a given disease within a given situation. In this context the term "optimal" refers to both the therapeutic suitability and the economic efficiency of a medication. While this task is impossible to accomplish manually in reasonable time, distributed information services could provide techniques to compare costs and therapeutic properties of pharmaceutical products automatically. For instance, the physician could select some brand drug and ask for any generics with the same composition and effects.

Within the scope of a hospital pharmacy two principal uses of pharmaceutical expert information may be distinguished. The main purpose is the drug dispension, usually performed automatically by a machine controlled by a special software system. Up to date this is the primary reason for electronic representation of medication information. The second task of the pharmacy is to check each medication for certain errors and inconsistencies, e.g., contraindications or serious interactions. This analysis, which is regular and mechanical in nature, is still done manually. Pharmacists sit in front of the screen and analyze medications using a book (BPI, 1993). There is a need for the automatization of medication analysis.

From the patient's point of view there is a need for information on the drugs he takes, on their usage and influence on his disease. This information has to be presented in a way understandable by the patient. This means that descriptions and instructions should not contain expert terminology and should not require a deeper understanding of the medical domain. This property of presentation has great influence on the patient's compliance. Apart from medical- or scientific-oriented scales, good patient information must be regarded as one component of a good therapy.

For any expert-oriented application it is desirable to access local and external documents which are relevant to a given situation or which explain some statement made by a system. For instance, if medication analysis reports a critical interaction, links should also be presented allowing the user to immediately access further information from standard literature or product documentation. This also suggests an application for medical students where a system can analyze their prescriptions and generate hints and advice.

PROBLEMS OF INFORMATION MANAGEMENT

Healthcare in general is characterized by high specialization and division of labour. Many professionals practice services for many patients, and data is produced at many places and may be needed at many others. Mobility of patients and the free choice of medical practitioners make this situation worse, as the individual's moves through the healthcare system leave trails consisting of distributed heterogeneous medical data. Thus, mobility of patients requires the mobility of data.

Pharmaceutical product information is distributed and heterogeneous in a similar way as is patient data. Furthermore, it is subject to fast change and reconditioning (Klein, Lee, Lei & Quereshi, 1996). It partly refers to patient medical data, as is with contraindications or age-specific dosage instructions. As a conclusion one may notice strong interconnections and dependencies between patient-specific medical data and pharmaceutical product information. Product documentation refers to patient-specific conditions, and patient therapy documentation refers to applied products.

The scattering and heterogenity of both patient history data and pharmaceutical product information lead to a variety of information problems for the different actors in the healthcare system. These problems refer to the availability of information in general and to the suitability of its representation in particular. Any of these role-specific information problems may finally be attributed to the two major problems of healthcare information management: representation and communication of expert and product information.

The *physician's information problem* refers to the treatment situation, where medications are composed and pharmaceutical products are compared. Far too much information would have to be taken into account in order to determine the most effective drugs regarding both therapeutic suitability and economic benefit. The problem consists of not having means to consult all of the available information sources manually, without the help of information technology (Klein et al., 1996).

The *pharmaceutical company's information problem* refers to both the communication of product information to the practitioners who prescribe it and the reflow of information gained through observation and experience. There are incentives for the industry to pass actual findings delivered by research and practice to the practitioners. There are also incentives to get informed on unknown drug effects.

The *insurance company's information problem* refers to the economic optimization of prescriptions through substitution of products. When taking care of the critical constraint of therapeutic equivalence, the substitution of brand products with generics bears a high potential of cost saving but effect keeping transformation. But this transformation requires knowledge and is desired to be performed automatically.

The *patient's information problem* refers to his situation at home, after being undeceived and instructed concerning drug application by the physician. In general, additional questions arise within this situation, and many aspects on disease, treatment, medication and instructions are not covered by the short and expensive consultation. From the patient's point of view there is a need to be able to get informed on these aspects whenever he wants to. From the physician's point of view there also is a need to delegate standard explanations to public media.

In the advent of the approaching information age, these information problems would not exist if there were reasonable means of applying information technology on the use of pharmaceutical expert information. This leads to the very question of information exchangeability within the healthcare sector regarding pharmaceutical products. As with any other kind of information on entities, the attributes of which are represented as references into certain external classifications, drug information may only be interpreted in regard to respective semantic reference systems.

There is one major conclusion to be drawn from the variety of role-specific information problems illustrated above. They have in common the need for information retrieval and processing concerning pharmaceutical products as well as its preparation in a suitable external representation. It seems that the reason for missing availability of pharmaceutical information is not its absence but the different languages existing in healthcare. Starting from that today there are means to transport information from one site to another, one may realize that the problem cannot be the absence of communication infrastructures. The answer is much more simple and complex at the same time: The problem lays within the information itself. The most serious problem blocking global communication in healthcare results from the lack of common languages.

MODELING PHARMACEUTICAL EXPERT INFORMATION

The formation of a universal information model for pharmaceutical products and their properties relates to domain specific coherencies concerning drugs and chemical substances as well as many aspects of software engineering in general. The former involves some understanding of the applications and effects of pharmaceuticals as well as the interdependencies among the different kinds of properties a drug may have. The latter involves design strategies and modeling techniques which meet today's requirements on compactness, reusability, flexibility and consistency of an internal representation (Riou, Pouliquen & Le Beux, 1999). Any pharmaceutical information model will be measured by its applicability in both fields.

The most fundamental awareness of any drug characteristics describing information model must be the distinction of two kinds of properties which may be regarded in different layers. These are properties which refer to a pharmaceutical product in isolation, on the one hand, and its behavior within the society of a group of drugs, on the other hand. The former is constant for a drug and will not change. The latter depends on the combination of drugs, the patient's state and many other aspects of its situation. This distinction may be captioned by the following assignments.

Static pharmaceutical information refers to drug properties which will not change when the drug's role within a medication or therapy changes. These properties include name, manufacturer, composition as well as obligatory information. Some of these properties may be represented as isolated data elements. This means that they are consistent on their own, and their interpretation is independent of certain external formalisms. Other properties (e.g., indications) are to be regarded as references into external semantic reference systems and may not be interpreted without the awareness of these formalisms. For instance, an indication can be regarded as a reference into an indication classification system. Of course one could model such properties as simple text fields. But in this case, only humans could understand this information.

Dynamic pharmaceutical information refers to drug characteristics concerning their behavior within a medication, especially the interplay of therapeutic effects. These characteristics are dynamic in nature because they may change with a drug's membership within different medications. They depend on various parameters of a clinical situation, including disease, patient demographic data, and, of course, other prescriptions. A drug may change the effect of other drugs and its effects may be changed by others.

A universal pharmaceutical information model should not refer to some specific classification system. Instead, it should refer to an abstract meta-structure which is capable of being instantiated by any classification. *Groupings*, as such meta-structures will be called in the following, are abstract trees representing inheritance hierarchies with implicit IS-A semantics. Inheritance is the principal property these structures provide: Any information associated with some group is also valid for each of its subgroups.

The universality of a pharmaceutical information model results from two major design properties. First, not one but a set of groupings reside in the knowledge base, providing a language for describing drug properties. Secondly, any drug may be assigned into not one but into any number of groups belonging to different groupings. The first design issue (multiple groupings) allows the consideration of different classification strategies and therefore the provision for different applications. Each grouping represents some specific view on the domain, focussing specific attributes of pharmaceutical products. The second design issue (multiple assignments) allows a maximum of precision concerning the representation of drug properties. One single assignment represents some feature of a drug. The entirety of assignments *characterizes* a drug (Wagner, 2000).

Groupings provide the vocabulary to specify relations between groups of drugs. *Conflicts* may be defined which relate one group to another and state that a combination of respective group members may cause a critical interaction. Additional conditions may be attached to such a conflict (e.g., indication, dosage) that further refine the set of cases where a conflict indicates an interaction. In this way, medication analysis may be reduced to the evaluation of group combinations.

Analytic processing capabilities of pharmaceutical expert information belong to the most beneficial properties a suitable information model can provide. Examples for their applications include automatic medication analysis, product comparison and even medication synthesis. There is no loss in functionality comparing to low-structured representations, because simple text documents may always be generated from precise representations. In addition, such a generation may take respect to the user's role and situation through inclusion and exclusion of information components, detail level determination and terminology selection.

The universality of a pharmaceutical information model based on the abstract grouping concept presented above may be illustrated by the following scenario. Assume a hospital pharmacy which is responsible for the daily validation of medication information. The pharmacy uses their own drug classification system which is tailored to the set of locally applied pharmaceuticals. The automation of medication analysis is desired on the condition that the local classification system needs not be replaced. This is the result of a requirements analysis made at Gummersbach.[2]

A medication analysis system based on the abstract grouping model meets these requirements. It suggests the following installation steps. First, the local classification system is entered into the system as a set of groups organized in a hierarchical structure. Secondly, the set of drugs is acquired focussing on static information and, thus, resulting in a linear list of records. Then, drugs are characterized by assigning one or more groups to each drug. Finally, conflicts are defined by selecting sets of groups the elements of which form a critical combination.

Then, medication analysis can be performed automatically. The system may be configured to let pass standard medications and to identify occurrences of critical drug combinations. In these cases the system may notify the users and the medication can be reviewed manually. From the pharmacist's point of view, automatic medication analysis can significantly reduce the number of cases which need human attention, while the local classification system can be used to specify the electronic knowledge needed for medication analysis.

Figure 1 illustrates the interplay between the different layers.

Figure 1: Interplay between objects, groups and conflicts

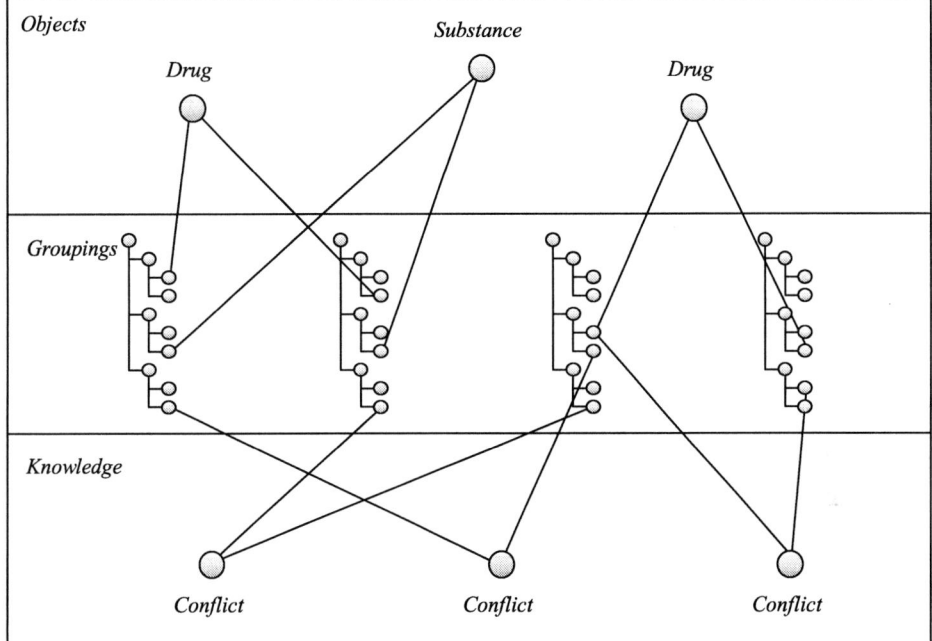

FURTHER RESEARCH

The formation of a universal pharmaceutical information model may suggest an internal representation within the context of pharmaceutical information systems, but it cannot provide solutions for problems resulting from the distinct roles, views and interests among the different actors in the healthcare system. Political agreement of common organization of expert information cannot be a concern of medical informatics.

The *unification problem* of healthcare refers to the chasm between universality and specifity of semantic reference systems. A universal classification system will not be suitable for specific applications and a classification which is suitable for a specific application will not be suitable for others. There seems to be no way out: The unification of semantic reference systems appears to be the only solution. This means that only the transformation between different classifications may enable communication of pharmaceutical expert knowledge.

The *standardization problem* refers to the development of semantic reference systems and to the institutionalization of their coordination. Starting from that diversity exists between different applications, roles and interests, it would be desirable to have these differences controlled by a global institution. The question is: Who will act as a global standards defining institution and which interests will be covered?

The major problem of internal drug representation is the formation of a universal information model. This in turn leads to the challenge of developing one universal drug classification system. Since any classification is a direct result of respective design strategies and focused applications, a universal drug classification system must be the result of common design strategies and common goals. The question is: What should be the major design strategy regarding a universal drug classification system?

The universality of scope claimed by uniformity of representation leads to the property of suitability. A universal classification system is exhaustive with regard to the focused domain, and in all probability it is oversized for small applications. This holds for monoaxial classifications. There is no proof for the theoretical existence of a universal (possibly multiaxial) classification which meets suitability of all applications. The question is: Can a universal pharmaceutical classification system meet the needs of specialized applications?

In the past many innovative concepts were developed to help overcome information problems in healthcare. Many projects failed at the requirements of data security. The latter has emerged to a political rather than a technical barrier of healthcare communication. In many cases, data security requirements for technical procedures exceed the requirements for equivalent manual procedures. For instance, the German electronic health insurance card was originally intended to store medical data on a patient's history. Today these cards do not contain any medical data. Under full agreement to the importance of data security, the question is: Will data security issues continue to block information processing and communication in healthcare?

Taking the diversity and heterogenity of pharmaceutical classification systems for granted and immutable, one possibility remains. This is the automated mapping of semantic references between different classification systems. This means that a reference into some classification may be automatically translated into a reference into another classification. Since almost any pair of classification trees consists of similar layers, such a translation is always possible. However, information loss is the price for this automated mapping when differences between classifications are too large.

DISCUSSION

The two-sided challenge of representation and communication of pharmaceutical product information illustrates the interdependency between information modeling and distribution in healthcare. Both dimensions affect the application of expert information. A pharmaceutical information model may be suitable for some applications and it may be suitable for communication. But it seems to be very difficult to design a model which is suitable regarding both dimensions.

As with any kind of information which is used in different application contexts, pharmaceutical product information may be structured in many ways, resulting in different and in most cases incompatible reference systems. Pharmaceuticals may be classified by indication or contra-indication, their composition of chemical substances, their pharmacological effects or therapeutic use. Any of these characteristics implicates certain applications of the resulting classification. In general, a classification system incorporates an application-specific strategy of decomposition. Strong interdependencies exist among properties like universality, complexity and suitability.

These circumstances suggest doubts on the theoretical existence of a universal classification system for a knowledge domain covering all application scenarios and meeting all requirements. These doubts get stronger under the consideration of suitability, which correlates extent and complexity of a classification system with extent and complexity of an application scenario. For instance, the German drug classification system provided by the "Rote Liste" covers about 9,000 commercial pharmaceutical products (BPI, 1993), while small hospital pharmacies deal with only about 800 drugs, which in their combination cover all medications. It seems that a universal classification system for a knowledge domain, which is exhaustive in nature, must be oversized for the needs of small applications (Wagner, 2000).

A scientific point of view enlightens many aspects of domain-specific classification development which identify the real problems predominantly within the expert domain itself. Many applications of classification systems make use of their hierarchical structure and object-oriented nature, if given. In fact, properties of classifications are applied in terms of inheritance and polymorphism. There exist reasonable doubts on the awareness of the object-oriented modeling paradigm in the community of classification developing medics. This can be proved with several classifications developed in medicine (Graubner, 1995), which in general match the requirements of knowledge representations, but fail at the vital design properties (e.g., ICD, ICD-O, ATC; Schulz, Zaiss, Brunner, Spinner & Klar, 1998).

CONCLUSION

Pharmaceutical expert information is characterized by steady increase in extent and complexity as well as by strong interweavement and penetration with interdependent references. These properties suggest the utilization of modern hypermedia technology, organizing knowledge as a network of linked pieces, enabling more intuitive, faster and easier navigation within the growing information space (Bush, 1988). There is a need for a suitable, role-oriented external representation of pharmaceutical expert information.

The application of pharmaceutical expert information when analyzing medications for contraindications and interactions is mainly mechanical in nature. It requires the ability to follow references and to compare codes. The electronic availability of medication information, which is often forced by drug dispensing systems, is not exhausted when these procedures are performed manually at the screen. There is a need for automatization of these procedures. This automatization requires a suitable internal representation of pharmaceutical expert information.

The universality of pharmaceutical information systems leaves open their conceptual environment. The different kinds of application scenarios pose strong requirements on the formalization of expert information. A universal representation has to be found which captures the nature of pharmaceuticals including common properties of drug classification systems. The diversity of the latter within the different healthcare organisations is to be regarded as the most serious barrier aggravating global communication within the healthcare system.

ENDNOTES

[1] This chapter is based on a paper titled "Representation and Communication of Pharmaceutical Expert Information" published in the Proceedings of the 8th European Conference on Information Systems, Vienna, Austria, July 2000, pages 1302-1308

[2] Kreiskrankenhaus Gummersbach GmbH Wilhelm-Breckow-Allee 20, 51643 Gummersbach, Germany

REFERENCES

Bundesverband der Pharmazeutischen Industrie. (1993). *Rote Liste 1993*. Arzneimittelverzeichnis des BPI. Aulendorf, Württ.: Editio Cantor Verlag.

Bush, Vannevar (1988). As we may think. *The Atlantic Monthly*, 1948 Reprinted in Computer-Supported Cooperative Work. In Greif, I. (Ed.). *A Book of Readings*. San Mateo. California: Morgan-Kaufmann.

Graubner, B. (1995). Wesentliche Klassifikationen für die medizinische Dokumentation in Deutschland und ihr Entwicklungsstand; in: ICIDH-*International Classification of Impairments, Disabilities and Handicaps*, Wiesbaden: Ullstein Mosby .

Hempel, L., Gräfe, A. K., Helmecke, D. and Schuchmann, H. W. (1994). *Arzneimittelinformationssysteme: Mehr Zeit für Wesentliches*. ABDATA Pharma-Daten-Service; Sonderdruck aus Pharmazeutische Zeitung, Nr. 41, 139. Jahrgang; S. 9-28, 88, 89.

Klein, S., Lee, R., Lei, L. and Quereshi, S. (1996). Pharmatica-Supporting the complex pharmaceutical information needs in the changing healthcare sector. *Electronic Markets*, 2.

Riou, C., Pouliquen, B. and Le Beux, P. (1999): A computer-assisted drug prescription system: The model and its implementation in the ATM knowledge base. In *Methods of Information in Medicine*, Department of Medical Informatics, Faculté de Médecine, Rennes, France, Schattauer, 38, 25-30.

Schulz, S., Zaiss, A., Brunner, R., Spinner, D. and Klar, R. (1998). Conversion problems concerning automated mapping from ICD-10 to ICD-9. In *Methods of Information in Medicine*, 37, 254-259.

Wagner, M. (2000). *Pharmazeutische Informationssysteme-Modellierung, Informationsstrukturen und Kommunikation Interdisziplinär Ausgerichteter Datenbanksysteme*, Vieweg; Wiesbaden.

Zagermann, P. (1996). ABDA-Datenbank: Einsatz der interaktionsdatei. In *Pharmazeutische Zeitung*, Govi-Verlag GmbH, Eschborn/Ts.; 141. Jahrgang; S. 2672-2675

Chapter VI

A Semantically Advanced Querying Methodology for Medical Knowledge and Decision Support

Epaminondas Kapetanios
Swiss Federal Institute of Technology, Switzerland

INTRODUCTION

A large part of all activities in healthcare deals with decision making regarding which examinations and tests need to be done or, on the basis of earlier examinations, which further tests need to be ordered. Recently, guidelines for an *appropriateness* and *necessity* indication of medical interventions have been elaborated and consulted in order to evaluate the quality of decisions in specific medical domains such as *cardiology* and *hysterectomy*.

In all decision supporting procedures, however, two types of knowledge are involved:

*   *Scientific or formal knowledge* which results from the literature such as books or articles in journals. This type of knowledge deals with *cognition* or *deduction*, which means that one must know and understand the principles of biological processes and relationships between pathophysiological conditions and disease symptoms.

- *Experiential knowledge* as condensed in well-documented patient databases or validated guidelines as specified by panels of experts. This type
of knowledge is related to *recognition* or *induction*, which means that
the clinician has seen certain symptoms before and recognizes the underlying disease.

In practice, these two types of knowledge are extremes of a continuum of
clinical knowledge and are interwoven when clinicians reason about signs and
symptoms of a specific patient. The same holds for quality assessment of
medical interventions. In most instances, clinicians have enough knowledge and
sufficient patient data are normally available to make the right decision. Yet,
there remain reasons why computers may be required such as:

1) Clinicians cannot keep up with the ever-increasing medical knowledge,
2) Healthcare organizations may mandate certain clinical practices both to
improve the quality of care and to lower the cost of care.

Computers require, however, that such knowledge should be structured
and formalized in order to make it available at the place and time when it is
needed. This is similar to data structuring in database management. Only then
can patient data and medical knowledge be used for computer-supported
decision making, assisting the knowledge in and the reasoning by human brains.

Major issues of medical knowledge which hamper computer-supported
decision making in healthcare, however, have been:

1) The ever-expanding knowledge,
2) That patient data are sometimes only partly available,
3) The problem of a specific patient may be new and unique.

Nevertheless, once medical knowledge for decision support has been
made available, the problems of structuring knowledge in a computer as well
as the availability of the right answers to clinicians arise. The latter is considered
as one of the crucial factors for dissemination and usage of medical knowledge,
since the query answering systems providing answers out of the structured
medical knowledge need to cope with scientific terms and conditions, which
underlie the specification of the medical domain.

Regardless how medical knowledge has been structured and managed,
clinicians, who are called *end users* in the following, must be mostly familiar
with the syntax formalisms of database or knowledge base specific query
languages (Catell & Barry, 1997; Groff & Weinberg, 1999; Maier, 1998;
Robie, Lapp & Schach, 1998) and/or understand very well the data/knowl-
edge model of the knowledge repository. In order to improve usability and
alleviate end users from syntax-based formulation of queries, advanced (visual)
query interfaces and/or languages (Catarci, Costabile, Levialdi & Batini, 1997)
have been designed, which provide a syntax-free (visual) formalism for query

formulation. The latter mostly relies on graphical presentations of conceptual models, which turn out to be very tedious to handle and difficult to understand, especially when large knowledge repositories and/or complex data schemes are concerned.

In all these systems, however, the role of *semantics* defined as the relationship of linguistic symbols and their meaning (Kromrey, 1994) as well as a constraints-based usage of scientific terms and results has not been the case. Furthermore, in many cases the end user does not really know what is relevant to her/him or how to ask for or to make use of the available knowledge in order to formulate a query, unless she/he fully understands it, i.e., sees, interpreting it. Typical query answering systems rely only on low-level features as expressed by the data/knowledge model on which the data/knowledge repository relies.

In this sense, the *semantically advanced query answering system* presented in this chapter changes the perspective of end user oriented query construction in that the *system guides the end user* to the construction of a reasonable query through simple visual query interfaces. They are used as an interaction platform or communication blackboard between end user and machine. The interaction mode between system and end user is conceived as an *abstract machine*, which receives as input the current query context (current state) and produces a semantically meaningful output (subsequent state). The decision on which output to produce strongly relies on the *semantics/meaning* of the terms constituting the vocabulary to be used for knowledge querying.

At the core of the query answering system resides **MDDQL** (**M**eaning **D**riven **D**ata **Q**uery **L**anguage; Kapetanios, Norrie & Fuhrer-Stakic, 2000), as visual query language through which there is no need to learn particular query language syntax and/or understand the semantics of the implemented data/knowledge model. **MDDQL** also enables, to some extent, the semantics-based guidance of the end user to the construction of analytical queries. Furthermore, in multilingual user communities, it is possible to construct a query with terms from more than one natural language with any effects on the query results. This is due to the separation between the world of symbols as used for the representation and structuring of knowledge and the world of multilingual vocabularies as used for the formulation of queries.

In order to illustrate the approach, an example from a current application of the system will be presented in this chapter as referring to the querying of medical knowledge for quality assessment in cardiological and gynecological decisions (Schilling et al., 2000). In particular, a *patient database* and a *knowledge base* are queried in the above medical domains in order to provide rather a *critiquing system* than *decision support* on its own. The *knowledge base* is constituted by a large set of classification rules concerning empirically estimated *appropriateness* and/or *necessity* of medical decisions. The rules perform inductions based on patients' symptoms and *symbolic models* of reasoning using logical operations, i.e., *qualitative* decision support. On the other hand, the *patient database* is queried in order to pose general queries concerning data from patients' records as well as validation of the *knowledge base*.

Organization: The chapter is organized as follows. Section 2 gives an overview of the related work as far as querying approaches and/or languages are concerned, especially those dealing with high-level querying. Section 3 describes a querying paradigm as refers to the medical domains of coronary angiography and revascularization. Section 4 gives an overview of the system in terms of the components–terminology base, inference engine, visual query interface, and query mediator–which are needed for the construction of a query. These components are part of **MDDQL** as a visual query language. Sections 5 and 6 give a more detailed description of the terminology base as a knowledge space and the inference engine, respectively. An overview of the **MDDQL** expressiveness is given in section 7. This is done in terms of the constructed semantic query trees to be submitted for transformation or execution. Section 8 discusses future trends and, finally, section 9 provides a conclusion.

BACKGROUND

At this point, we embark on an overview of traditional and proposed query languages to be used as a means of addressing instances in a database through the specification of conditions rather than simply using some browsing techniques as happens to be the case in meta-data databases or knowledge/ontology browsers. We mainly distinguish between *instance description oriented query languages* and *navigation style oriented* ones.

Instance Description Oriented Query Languages

As stated previously, this family of query languages (Abiteboul, Hull & Vianu, 1995) deals with instances, which are given on the basis of the specification of their properties. The family consists of query languages based 1) on algebraic operations (selection, projection, union, difference, etc.) as known from the *relational* or *collection* algebra, 2) on logic with an expressive power equivalent to the algebra such as *relational calculus*, and 3) on logic programming such as *datalog* which can be viewed as logic programming without function symbols.

They are all *set-at-a-time* oriented, in the sense that they focus on identifying and uniformly manipulating sets of instances rather than identifying instances individually and using loops to manipulate groups of instances. However, since instances in the answer are specified by properties they satisfy, with no reference to the algorithm producing them, *relational calculus* and *datalog* are conceptually *declarative*, whereas *algebra-based languages* are conceptually *procedural*.

Representative query languages of this family are SQL (Structured Query Language; Groff & Weinberg, 1999), OQL (Object-oriented Query Language; Catell & Barry, 1997; Kifer, Kim & Sagiv, 1992) for relational and object-oriented databases respectively, as well as the query languages which have been proposed for XML based Web documents, such as XML-QL (Maier, 1998), Lorel (Abiteboul, Quass, McHugh, Widom & Wiener, 1997), XQL (XML Query Language; Robie, Lapp & Schach, 1998) and XSL (eXtensible Stylesheet Language; Group, 1998).

In particular, *SQL/OQL*-like query languages are suitable for well-structured data and presuppose a full understanding of the data model semantics. Furthermore, natural language based interpretation of model elements such as classes, relations, attributes and values as well as operational semantics are not taken into account during query formulation. Semantic consistency within a query is also not a matter of subject. Therefore, they mostly meet the requirements of application programmers but not those of *end users*.

The same holds for XML-QL, Lorel, XSL and XQL. *XML-QL* has been designed at AT&T Labs as part of the Strudel project. The language extends SQL with an explicit CONSTRUCT clause for building the document resulting from the query and uses the *element patterns* (patterns built on top of XML syntax) to match data in an XML document. XML-QL can express queries as well as transformations for integrating XML data from different sources. *Lorel* was originally designed for querying semi-structured data and has now been extended to XML data. It was conceived and implemented at Stanford University and has a user-friendly SQL/OQL style including a strong mechanism for type coercion and permits very powerful path expressions which are extremely useful when the structure of the documents is not known in advance.

XSL consists of a collection of template rules. Each template rule has two parts: a pattern which is matched against nodes in the source tree and a template which is instantiated to form part of the result tree. XSL makes use of the expression language, defined by XPath (Clark, 1999), for selecting elements for processing, for conditional processing and for generating text. *XQL* can be considered a natural extension to the XSL pattern syntax. It has been designed with the goal of being syntactically very simple and compact (a query could be part of a URL) with a reduced expressive power.

Navigation Style Oriented Querying

In the following, we are dealing with the second major family of query languages which adopts a navigational style for the construction of a query or addressing a particular query result. They are mainly classified into two categories: a) those which are enriched with visual query interfaces or metaphors and b) those which are based on graph formalisms and/or traversal of paths/links.

Visual Query Systems and Languages

The importance of avoiding an underlying query language formalism when end users need to pose queries to a database system has received much attention during the last 10-15 years within the database research community (Berztiss, 1993; Cardiff, Catarci & Santucci, 1997; Clark & Wu, 1994; Florescu, Raschid & Valduriez, 1996; Haw, Goble & Rector, 1994; Meoevoli, Rafanelli & Ricci, 1994; Merz & King, 1994; Ozsoyoglou & Wang, 1993; Papantonakis & King, 1995; Siau, Chan & Tan, 1992) and many *visual query systems* (VQS) and/or *visual query languages* (VQL) have been developed to alleviate the end users' tasks. A survey of these approaches is given in Catarci et al. (1997). A classification of VQSs based on the criteria of expressive power, usability and categories of potential users is given in Batini, Catarci, Costabile & Levialdi (1991).

Despite the fact that VQSs can be seen as an evolution of query languages adopted in database management systems in order to improve the effectiveness of human-computer interaction (HCI; Chan, Siau & Wei, 1998), the notion of *usability* as defined in Batini et al. (1991) was mainly specified in terms of the models used in VQSs for denoting both data and queries, their corresponding visual representations and the strategies provided by the system in order to formulate a query.

This definition does not reflect the human-computer interaction (HCI) view of usability as a software quality related to user perception and fruition of the software system. In particular, the data model has no significant impact on the user perception of the system. VQSs mainly rely on the integration of the data model and query language in a user-database interface (Chan, 1997) as well as presentation and interaction components that together form a graphical user interface (Murray, Goble & Paton, 1998).

On the other hand, a classification of VQSs based on the criteria of *visual representation* of queries and query results is given in Catarci et al. (1997)

where VQSs are mainly classified into *form-based*, *diagrammatic* and *iconic,* according to the representation of the domain of interest for query formulation and/or the query result. They all refer to information repositories dealing with alphanumeric data and not with semi-structured, video or audio data.

In particular, form-based representation techniques became popular since the appearance of the first example-based query language *QBE* (Query-By-Example) in 1977 (Zloof, 1977). More than a dozen such languages have been proposed and/or implemented (Epstein, 1991; Houben & Paredaens, 1989; Shirota, Shirai & Kunii, 1989; Wegner, 1989). They are designed for different application domains such as statistical and scientific applications, office automation, historical databases and spatial databases and have different capabilities and expressive powers. Most example-based languages are based on revised versions of Codd's domain relational calculus.

A survey of example-based query languages is given in Ozsoyoglou & Wang (1993), which compares the features of 12 example-based languages in terms of (1) query specification and interpretation, (2) object manipulation, (3) query language constructs, and (4) query processing techniques. Example-based query languages allow users to specify queries through an example that is constructed graphically, thus utilizing the analogy between a semantically meaningful example and the query. These languages provided a user-friendly graphical interface, especially for relational databases.

At the end of the '80s and beginning of the '90s, we observe a shift of visual querying towards entity-relationship (ER) based diagrammatic representation of database schemata such as Angelaccio, Catarci and Santucci, (1990), Dennebouy et al. (1995), and Elmarsi and Wiederhold (1981). They all rely on typical query operators such as selection of visual elements, traversal on adjacent element and creation of a bridge between disconnected elements. In some systems like Dennebouy et al. (1995) and Elmarsi and Wiederhold (1981), the navigation path is implicitly expressed by transforming the query schema into a tree where nodes appear in the selected order. Some systems also allow the users to define their own visualization rather than creating tailored displays of data. A proof of equivalence of navigations in Angelaccio, Catarci and Santucci (1990) to relational algebra is given in Catarci and Santucci (1988).

In contrast with Entity-Relationship based diagrams, the iconic query language *QBI* (Query-By-Icon; Massari, Pavani, Saladini & Chrysanthis, 1995) is not concerned with path expressions but it allows users to query and understand the content of a database by manipulating icons. It provides intentional browsing through meta-query tools that assist in the formulation of complete queries in an incremental manner by using icons. A comparison study between using "diagrams" and using "icons" for the formulation of queries is given in Badre, Catarci, Massari and Santucci (1996). QBI is based on a slight variation of the Graph Model introduced in Catarci, Santucci and Angelaccio (1993).

Hybrid VQSs have also been proposed which offer an arbitrary combination of the above three visual formalisms, either offering the user various alternative representations of databases and queries, or combining different visual formalisms into a single representation. The major categories have been those VQSs which combine forms and diagrams (King & Melville, 1984), diagrams and icons (Groette & Nilsson, 1988), and form, diagrams and icons (Cinque, Levialdi & Ferloni, 1991).

Although it was reported that manipulating diagrams is definitely easier than writing SQL commands, it was first an Esprit project, *VENUS*, which started addressing the real "meaning" of usability by having users such as hospitals and research centers. The efficiency of processing queries in the VENUS environment is considered in Cardiff et al. (1997). The goal of this system was to allow a user to query multiple data sources by interacting with a conceptually single database and by providing a query interface based on the use of visual formalisms. This system takes advantage of existing multi-database (MDB) query processing strategies and deals with optimization issues by considering the exploitation of inter- and intra-schema semantics. A methodology is presented which harnesses knowledge of the set-theoretic relationships between classes participating in a query, and demonstrates that significant saving can be made in MDB query processing.

Aiming at a different category of users, the *Vista* visual language (Meoevoli et al., 1994) has been proposed, which can be easily used by users when they interact with a statistical database in order to manipulate data directly. This language uses a directed acyclic graph as an internal model for the database scheme representation. The operators on the statistical data manipulate their descriptive elements, while the statistical processing is performed through statistical packages. This visual language permits querying in the database scheme and the selection of subschemes. Vista is integrated in the statistical database management system Adams (Aggregate Data Management System). This system allows also an interaction through a keyword-based language. The main tasks performed by the Adams system are the graphical visualization of the database scheme, the direct manipulation and the expression of query commands.

Besides VENUS, further visual query languages have also been proposed and applied to multi-database query facilities. The subject of Merz and King

(1994) was the architecture and design of a multi-database query facility. These databases contain structured data, typical for business applications. Problems addressed are: presenting a uniform interface for retrieving data from multiple databases, providing autonomy for the component databases, and defining an architecture for semantic services. *DIRECT* is a query facility for heterogeneous databases. The databases and their definitions can differ in their data models, names, types and encoded values. Instead of creating a global schema, descriptions of different databases are allowed to coexist. DIRECT has been exercised with operational databases that are part of an automated business system.

Clark and Wu (1994) proposed the query language *DFQL* (Data Flow Query Language), which has been designed to mitigate SQL's problems concerning its syntax formalism. DFQL provides a graphical interface based on the data-flow paradigm in order to allow a user to construct queries easily and incrementally for a relational database. DFQL is relationally complete, maintains relational operational closure, and is designed to be easily extensible by the end user. A prototype DFQL system has been implemented.

The query language *Vizla* (Berztiss, 1993) builds up answers to queries by pointing to representations of sets and functions in a conceptual model of the database of an application, and to iconic identifiers of computational operators or control constructs. The primary use of Vizla is in the validation of conceptual models of information systems, but it is to be developed into a user interface to a prototyping language for information and control systems. Moreover, it can be regarded as a visual programming language in its own right. As such it is based on abstract data types.

However, one of the major lessons learned was the understanding that users had many difficulties even if ER-based diagrammatic presentations of database schemes were chosen as a query interface–the well-known "wires and meshes" problem. Another major difficulty users had was the combination of selection conditions on attributes using the Boolean connectives AND, OR, NOT. Moreover, users locked up in an option. By then, developers suspected that being a database expert was not enough to design "interactive" information systems.

Furthermore, the conceptual or implementation–mostly relational–model is mainly integrated into the query paradigm. Query construction takes place in terms of navigation through a graphical representation of the entire model and without any consideration of *meaning* in terms of semantic constraints holding among query elements, i.e., mutually exclusive properties and/or values, as well as their interpretations. Moreover, usage of different natural languages has not been the case.

Even if VQSs are characterized from the perspective of end user/system interaction strategies where mainly top-down, browsing and schema simplification strategies are considered for understanding the domain of interest prior to formulation of a query, *meaning* is not taken into consideration as a predominant feature of the interaction strategy. Additionally, understanding of domain of interest and formulation of a query have been conceived as two different processes.

Graph-Based Formalisms and Traversal Paths/Links

Graph-based formalisms have been proposed for both object-oriented DBMSs where traversal paths can be expressed as queries (Chavda & Wood, 1997; Yu & Meng, 1998) and for Web query systems (Ceri, Comai, Damiani & Fraternali, 1999; Li & Shim, 1998). In particular, Chavda and Wood (1997) and Yu and Meng (1998) proposed querying object-oriented databases by following paths or links connecting objects to each other. Li and Shim (1998) propose a visual user interface–*WebIFQ* (Web In-Frame-Query)–which assists users in specifying queries and visualizing query criteria by including document meta-data, structures and linkage information.

XML-GL (Ceri et al., 1999) is considered as a *graphical query interface to XML*, playing the same role as graphical query interfaces such as Query-by-Example for the relational world presented above. However, since query formulation is done by means of *labeled XML graphs*, we classified XML-GL into the family of graph-based formalisms. The language has been designed at Politecnico di Milano and the implementation is ongoing. The basic idea is the usage of a graphical representation of XML documents and DTDs (document type definitions) by means of labelled graphs.

Traversal like approaches using graph queries or high level concepts also underlie the development of query interfaces for large clinical databases (Banhart & Klaeren, 1995; Hripcsak, Allen, Cimino & Lee, 1996; Liepins, Curran, Renshaw & Maisey, 1998; Taira et al., 1996), or for general purpose systems (Chu et al., 1996; Chu, Merzbacher & Berkovich, 1993; Doan, Paton, Kilgour & Qaimari, 1995; Gil, Gray & Kemp, 1999; Murray, Goble & Paton, 1998; Murray, Paton & Goble, 1998; Zhang, Chu, Meng & Kong, 1999). The query generators mostly use an object-oriented data model or functional models such as in case of Gil et al. (1999).

Despite the fact that in all these approaches query construction is done by navigational issues where the end user does not need to learn a particular query language, it is very often hard to operate on complex or large diagrammatic representations, which occurs when large database schemes are considered. Furthermore, semantic constraints within a query are not taken into account, which might lead to semantically incorrect queries.

Even in cases such as Gil et al. (1999) and Zhang et al. (1999), where data values are considered for the incremental formulation of the final query, these values are not addressed within well-restricted value domains and cannot be the subject of a semantically meaningful consideration of values for conditional statements given the current query context. Without such a meaningful consideration of values, queries might be constructed the results of which are, semantically speaking, *not worth addressing* and might lead to expensive operations without considerable results.

In Gil et al. (1999), where values can be taken from intermediate results during query construction and further refined to form the final query, we are still faced with the problem of addressing values from pre-calculated results which, in case of large databases, might exceed several hundreds of rows or tuples.

A QUERYING PARADIGM

In order to illustrate the philosophy behind the presented querying approach, a querying paradigm is described in the following which refers to querying of a medical knowledge base (experiential knowledge) as specified by a panel of experts. The knowledge base consists of a large set of classification rules which target at numerical scores of appropriateness and necessity of medical interventions in cardiology and gynecology. The classification rules are represented by decision (mapping) tables managed by a relational database engine.

For instance, consider the following classification rules (guidelines) as stored in decision tables. The first holds within the domain of interest *Patients candidate for Coronary Angiography* and the second within the domain of interest *Patients candidate for Coronary Revascularization*. The applicable appropriateness and necessity scores lie in the range of [1, 9] for a particular patient.

Rule CA-1:
IF patient has atypical angina AND
stress test is positive AND
gender is female AND
age is under 50
THEN appropriateness of coronary angiography is 7.0

Rule CR-1:
IF patient is asymptomatic AND

stress test is positive AND
disease is left main AND
ejection fraction is > 50 %
THEN appropriateness of revascularization is 7.0

In order to retrieve the score of appropriateness as related to CA-1 rule from a relational database, as always, one needs to write an SQL query with all those relevant conditions as refer to the class of the patient and the properties {*stress test, gender, age*} with their corresponding value restrictions. However, representation of a large number of attributes and values within a set of relations (tables) as a representational model of the knowledge base is bound to the meaningless structural properties of the underlying data model. This also holds for the cases where tree-like data models such as XML or object-oriented would have been preferred. In addition, the formulation of SQL or other database-specific queries, which include {*attribute, value*} pairs as querying conditions, presupposes great familiarity of the end user with query language syntax formalisms and complete understanding of the database schema semantics and data interpretation, which is cumbersome when large data repositories are affected. Even if navigational style oriented querying approaches are adopted, schema semantics and data interpretation still remain a challenging issue.

The querying methodology presented in this chapter overcomes these problems in that the end user is supported in formulating the appropriate query through an interactive query formulation technique. The constructed queries should, however, comply with the database semantics.

The question arising now is what should the system suggest and how should it present the relevant terms to the end user. In other words, which is the most appropriate interaction strategy for a query construction. For example, it makes sense to consider the attributes of *stress test*, *gender* and *age* only, if we are concerned with a patient which has been classified as having *atypical angina* and the given domain of interest is *coronary angiography*. Given another domain of interest such as *coronary revascularization*, another set of attributes and/or finite value domains will be relevant, even if the classification of a patient is the same.

In addition, since attributes and values can be presented in different natural languages or underie a specification, which is a common issue when domain science terminology is used, presentation of the suggested attributes or values is done in terms of words and not of the common reference symbols of the implementation. For instance, the words *risk factor* or *Risikofaktor*, standing for the description of an attribute, are mapped to the common reference symbol

RF.

Furthermore, the attributes and values are presented to the user with their *intentional meaning*, the absence of which during user-system interaction might lead to wrongly interpreted attributes and values and, consequently, to semantically wrong queries.

For example, the attribute *risk factors* underlies the definition

Risk factors for coronary artery disease are defined as: diabetes mellitus, hypertension, dyslipidemia, current smoking and previous smoking more than ten pack years, positive family history (e.g. history of a myocardial infarction, previous revascularization in a patient of less than 65 years, known coronary artery disease in patients less then 65 years) and male gender. For patients with atypical angina, male gender is not defined as a risk factor.

whereas the value *asymptomatic* underlies the definition

Asymptomatic are those patients with no history of atypical or chronic stable angina with normal physical activity; patients with a history of chronic stable angina who have had no angina on their current medical regimen for more than three months; patients who have sustained a myocardial infarction more than 21 days ago with no recurrence of angina, and patients screened because of risk factors or because of high risk occupation.

Trying to answer the query which is the appropriateness score for patients with atypical angina having stress test positive, gender female and age under 50, as related to the rule CA-1, with the interaction strategy of **MDDQL**, the system first suggests a set of terms standing for relevant patients' classes as depicted in Figure 1. The meaning of each term can be viewed before a decision is made for further consideration.

Having selected atypical angina as the relevant class of patient, the system further suggests the properties {*stress test, gender*} for instantiation, where each of these properties is assigned to corresponding value domains such as {*positive, negative, indeterminate, not done*} and {*male, female*}, respectively.

Subsequently, Figure 2 depicts the inferred suggestion of the system due to the current query state as given by the set of instantiation pairs {{*patient, atypical angina*}, {*stress test, positive*}, {*gender, female*}}. Since *age* depends on the instantiation of *gender*, instantiation of *age* will follow that of *gender*. Having selected the value *female* for *gender*, the inferred set of terms as finite value domain for *age* will be {*under 50, 50 to 75, 75 or older*}. Otherwise, if the value for *gender* would have been set to *male*, the value *under 50* would have been excluded from the inferred (suggested) set of terms as finite

Figure 1: Moving into the initial query state {patient, atypical angina}

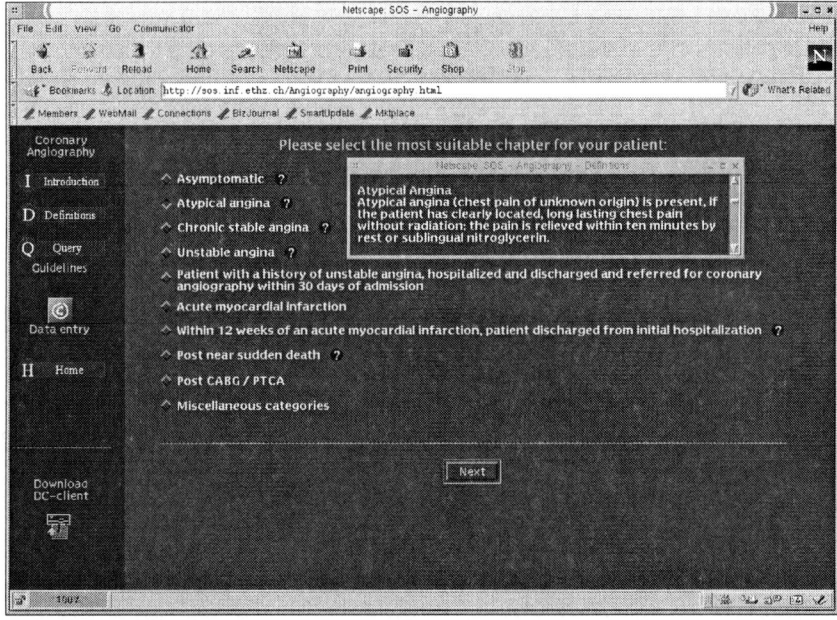

Figure 2: Potential extension of current query state

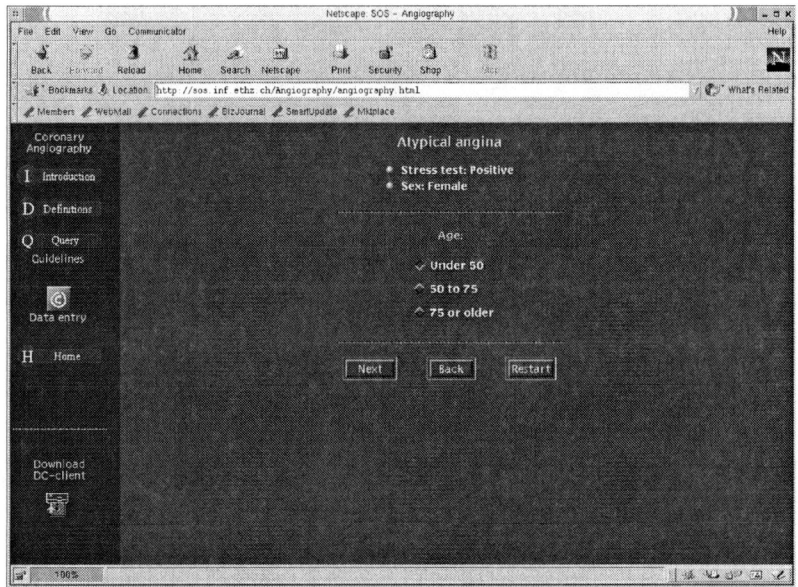

value domain, since the precondition *not (gender, male)* assigned to the term *under 50* is not satisfied.

All terms appear in some user-preferred natural language. Therefore, the same query could have been expressed in terms of the German language without affecting the query results, since the submitted query is translated into an SQL-statement in terms of the common reference symbols of the implementation model.

More advanced queries are also enabled which are targeted at the *patient database* and might include operations and/or comparison operators as assigned to query terms. They are also the subject of semantics-based inferences, since assignment of particular operations or operators is also a matter of query context. Therefore, *meaning* of terms is also extended to operational terms. We will return to this point in section 5. In the following, an overview of the system is given in terms of its components which are behind the scenes.

QUERY LANGUAGE SYSTEM COMPONENTS

Figure 3 gives an overview of the system components of the **MDDQL** query language. The system can be classified as a *client-sided, database-supported Web application*, which clearly considers database management systems (DBMSs) as data/knowledge repositories at the back-end layer. For example, both the *patient database* and the *knowledge base* (experiential knowledge) are provided by relational engines where the rules of the knowledge base are structured as decision tables. Further knowledge repositories to be considered at the back-end layer are XML-based document management systems for the representation of *scientific or formal knowledge* in terms of literature (books, articles).

Given the data/knowledge repositories at the back-end layer, the two other layers are related to the query language system components. They are provided by an application server and are classified as follows:

1. The terminology base as refers to the alphabet or vocabulary of a particular medical domain as well as the operational terms,
2. the inference engine which semantically guides the construction of a query,
3. the visual query interface,
4. the query mediator.

The component of query mediator receives the constructed **MDDQL** semantic query tree and transforms it to a database-specific query based on a transformation logic which makes use of the operations of the relational or collection algebra. The mapping between terms in the terminology base and

Figure 3: An overview of the MDDQL query language system components

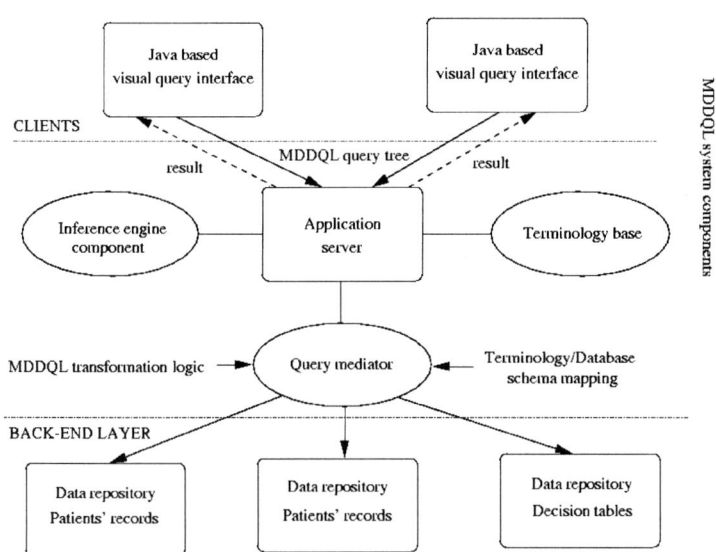

database schema or model semantics is also taken into consideration during query transformation. The mediation component outlies the scope of this chapter and, therefore, will not be further considered.

The inference engine relies upon the libraries with the inference algorithms as implemented in *Java*, which are needed for the semantics-based guidance of the end user for the construction of a query. In order to perform the inference tasks according to a given query context, the meaning of terms as well as semantic constraints, which are both represented in the terminology base, are taken into account. Both components will be described thoroughly in sections 5 and 6.

Since **MDDQL** enables the construction of queries in more than one natural language, additional components have been added to the middle layer which not only generate or transform an **MDDQL**-based query into a database-specific query language such as SQL or XML-QL but also translate the elements of the particular natural language to the symbols underlying the implementation model of the data repositories.

Currently, there are two representation platforms considered for the terminology base: a) **omsJava** (Kobler & Norrie, 1999; Norrie, 1993; Norrie, 1995), which is used as an object-oriented application framework for the *Java* programming language, and b) **XML**. The first relies on extensions of the data modeling constructs provided by the *Java* programming language through constructs supporting manipulation of binary associations.

Since the inference engine semantically drives the end user to the construction of a meaningful query, the visual query interface is implemented on the basis of an Application Programming Interface (API) as provided by the inference engine. Decisions, for example, which terms make sense to suggest and in which presentation form, are taken by the inference engine, whereas the visual query interfaces provide a visual formalism for the representation of the whole query.

The classification of the system as client-sided, database-supported Web application is dictated by the fact that the query construction process takes place at the client side having parts of the inference engine and the visual query interface downloaded as a packaged JAR file. This, in turn, must be installed and run on the client machines. The terminology base with the query vocabulary terms is available on the server-side part of the system (middle layer in Figure 3).

In the following, we will have an insight into the components of **MDDQL**, especially those first two components, which are responsible for the semantics-based inferences of terms during query construction. Implementation of the visual query interface will not be covered by this chapter.

STRUCTURE OF THE KNOWLEDGE SPACE OF TERMS

Since knowledge concerning semantics of medical domain vocabularies is extensively used by the inference engine of **MDDQL**, a model for the representation of these vocabularies is presented in this section. The major issue, however, is how to structure or form a knowledge space in order to enable inferences to be drawn during query construction. An overview of the structure of the knowledge space is given in Figure 4.

Given that the inference engine works on the basis of navigation through the structured knowledge space by looking for adjacent terms when a query context (set of terms) is already given, the knowledge space consists of various interconnected layers which are mainly classified into two categories: a) layers with all **MDDQL** *term instances* as the set I_A which consists of the set of *domain terms* (concept, property, value domain) and the set of *operational terms*, and b) layers with all *term classes* as the set C_T of all known classifications of I_A. Interconnecting links between I_A and C_T indicate the membership of term instances to particular classes. Formally speaking, the knowledge space I is structured as a *multilayered, directed acyclic graph– ML-DAG* defined as follows:

Definition: An ML-DAG (V, E) is a directed acyclic graph where V is the

Figure 4: Structure of the knowledge space of an MDDQL vocabulary

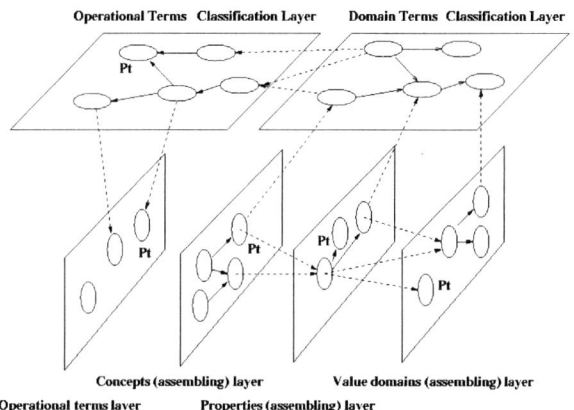

set of vertices $v_1, v_2, ..., v_n$, $n \in N$ and E the set of edges (v_{n-1}, v_n), v_{n-1}, $v_n \in V$. V is partitioned into layers $L_1, L_2, ..., L_m$, $m \in N$ such that $L_i \subset V$, $i = 1, ..., m$ and $L_i \cap L_j = \emptyset$ for any $i \neq j$. Additionally, $E_R \subset E$ is the set of *recursive links* and $E_I \subset E$ the set of *interconnecting links*, i.e., E_R consists of edges (v_i, v_j) where $v_i \in L_k$ and $v_j \in L_k$ and for E_I it holds that $v_i \in L_k$, $v_j \in L_p$ where $k \neq p$.

The part of the knowledge space which refers to the domain term instances and their classifications is application domain dependent and, therefore, becomes a matter of *specification of conceptualization* during requirements analysis. Despite the fact that the term *conceptual model* has been coined with many slight differences in meaning (Juristo & Moreno, 2000), the essential features have been representation and understanding of the problem raised by the users and/or customers in order to reach an agreement on the scope of the solution in terms of providing or building a software system that solves the problem in question. Therefore, the objective of conceptual models in the field of data engineering is to represent the domain of discourse in terms of data involved in the problem and operations on these data (Loukopoulos & Karakostas, 1995; Mylopoulos, Borgida, Jarke & Koubarakis, 1990; Wieringa, 1995).

Embedding meaning in vocabularies, however, poses additional challenges to conceptual modeling. Representation of *meaning* of terms is done in a twofold way: a) through the representation of *intentional meaning of terms* and b) through the *context of terms*. In particular, the intentional meaning of terms for both application domain terms and operational terms is expressed by having all terms conceived and represented as *objects*, and therefore, not symbols but symbols with additional properties are considered to be the elements of the vocabulary.

On the other hand, the context of terms is expressed a) by a connectionism model, i.e., set of edges in the knowledge space, which provides the basis for connecting/assigning terms to each other, and b) the expressions of constraints holding among them. The latter refers to constraints holding among application domain terms as well as between operational and domain terms.

Representation of Intentional Meaning of Terms

A term is conceived as a *unit of thought* and is assigned a *term unique identifier–TUI*. A TUI might refer to one or more natural language words. A word in a particular natural language such as English or German might be assigned to a term as representative of a class of synonyms but not as the identifier of the term. The set of all **MDDQL** terms is divided into two main categories: a) the set of application domain terms and b) the set of operational terms.

The Domain Terms

Intentional meaning of domain terms T_D is expressed by additional properties which might refer to:
• definitions and/or explanations of terms,
• underlying measurement units,
• images,
• synonyms, etc.

Therefore, domain terms are presented to the end user for selection during query construction together with the additional properties. Consequently, only well-understood domain terms are selected/included in the query. All domain terms are mainly classified either as *concepts*, *properties* or *values*.

In order to illustrate the approach, an example is given as taken from the medical domain of interest *hysterectomy*, for which the following subset of domain (alphabet) terms have been specified and expressed in *English*.

{tui = 1, name = 'Patient', symbol = 'P1'}

{tui = 40, name = 'Premenopausal', symbol = 'P2'}

{tui = 60, name = 'Abnormal uterine bleeding (no leiomyomata)', symbol = 'P43', description = 'Unacceptable social and/or hygienic situation for the patient.'}

{tui = 100, name = 'Major Impairment', symbol = 'MImpair', description = 'During the last 3 months, the patient stayed home (missed work or cancelled all activities) for 1 day per month because of pain or discomfort.'}

{tui = 200, name = 'Estimated uterine weight', symbol = 'EVW', description = 'Based on clinical experience estimated in grams (280grams corresponds to 12 weeks of gestation). If the patient is given GnRH agonists prior to proposed surgery, use size prior to treatment with these agent', measurementUnit = 'grams'}

{tui = 1000, name = '> 20 % within 6 months', symbol = '>20'}

{tui = 2000, name = '> 50 % within 6 months', symbol = '>50'}

{tui = 3000, name = 'Positive', symbol = 'PGC'}

{tui = 4000, name = 'Negative', symbol = 'NotPGC'}

{tui = 5000, name = 'Postmenopausal', symbol = 'Post', description = 'Amenorrhea greater or equal 12 months serum: FSH greater 30 IU/l estradiol less 100 pmol/l.'}

Since all alphabet terms are conceived as objects having assigned natural language words as value of the field *name*, there is a mapping to underlying, natural-language word-independent symbols as referred to by the value of the field *symbol*. Therefore, the structural properties remain the same when we change natural language reference.

Domain terms underlie a further classification structure which enables an insight into the nature of terms which is conceived as a contribution to the intentional meaning of terms. This is similar to operational terms as presented in the following. The nature of terms is further described by the names of classes or collections into which terms as information objects have been further classified. We will call the set of all classes participating in the representation model of meaning of terms C_T. For example, we might distinguish between *Categorical* or *Numerical Variables* as classes into which terms classified as *Properties* will be further classified.

The Operational Terms

Similar to domain terms, operational terms are also considered as objects having intentional meaning represented by linguistic elements, explanations stating the purpose of application, etc. Operational terms are mainly classified as *comparison operators* such as

{tui = OP4, name = 'equal to or more than', symbol = '>'}

or *univariate functions* such as descriptive statistical operations

{tui = OP17, name = 'Maximum', symbol = 'MAX'}_

{tui = OP21, name = 'Median Value', symbol = 'CV',

explanation = 'The value under which fifty percent of all values lie.'}

which are usually applied for the summary description of data concerning a single variable (property term).

Further functions for the study of bivariate or multivariate data are also considered when one wishes to discover relationships which might exist between the variables (property terms), or how strong the relationships appear to be, and whether one variable of primary interest can be effectively predicted from information on the values of the other variables.

Bivariate functions operate over two variables such as *Gender* and *Type of occupation* of patients. Typical examples of such operational terms are:
{tui = OP100, name = 'Cross-classification', symbol = 'CC',
explanation = 'Tabulation of data related to two variables.'}
{tui = OP101, name = 'Scatter diagram', symbol = 'SD',
explanation = 'Study of the relationship between two variables.'}

Operational terms constitute together with application domain terms the terminology base which is part of the query language system as depicted in Figure 3. The inference engine is also responsible for a semantically correct assignment of operations and/or operators to query terms with respect to a given query context.

Representation of Context of Terms

The Connectionism of Knowledge Space

The connectionism model supports two major categories of links holding among the objects which constitute the knowledge space: a) those (interconnecting) links, e.g., connecting *concepts* to *properties* and *properties* to *values* by forming well-restricted value domains, and b) those (recursive) links connecting terms of the same class to each other. Links also hold within C_T as well as between I_A and C_T.

Considerations for the set of domain alphabet terms I_D: At this point, we mainly refer to the objects which belong to the concepts, properties and values (assembling) layers (see also Figure 4). Within these layers, the set of interconnecting links E_I can be defined in database theoretic terms as follows. We assume that there exists a countably finite set of attributes *att*, i.e., domain terms classified as *properties*, and a countably finite set of values *dom*, disjoint from *att*, i.e., domain terms classified as *value terms*.

The interconnecting links holding between the layers of *property terms* and *value terms* establish a mapping *Dom* on *att* such that *Dom(A)* is a set called the *domain* of $A \in att$. Moreover, the interconnecting links holding between *concept terms* and *property terms* establish a mapping *Pr* on the countably finite set *con* of domain terms standing for *concepts* such that *Pr(C)*, $C \in con$, is a set called the *characteristic properties* of *C*.

In addition, *recursive links* enable the refinement of *concepts*, *properties*, and/or *values*, respectively. Therefore, it is possible to refer to the *reflexive mappings R* on *con, att, dom,* respectively, such that *R(con)* is a set of *constituent concepts, R(att)* is the set of *constituent properties, R(dom)* is the set of *constituent values.* Therefore, the *power sets P(Dom(A))* and *P(A(C))* are also allowed as a range of the mappings *Dom* and *A,* respectively. This enables the consideration of recursively structured (multidimensional) properties as well as recursively structured value domains.

Given that $N:M$ mappings are allowed for both kinds of links, it is possible to assign the same *property* to more than one *concept* as well as the same *value* to more than one *property.* The same holds for the *recursive links,* e.g., a *concept* or *property* might be connected to more than one *complex concept* or *property.* Thus a particular term might participate in the refinement of more than one complex term.

Definition: An *assembled* concept, property or value term is a *proper path* $p_a = v_0, ..., v_n, n \in N$ where $v_i, 0 \leq i \leq n,$ belong to the same layer L_m with $m \in \{c, p, v\}$ and all holding connecting edges belong to $E_R \subset E.$

For instance, the concept terms with term unique identifiers $\{1, 40, 60\}$ are connected with recursive links within $L_c,$ which, if they are followed, form the *assembled concept {patient, premenopausal, abnormal utcrine bleeding (no leiomyomata)}* to participate in a query. Moreover, since the underlying model is a directed acyclic graph, both assembled concepts *{patient, premenopausal, fear of cancer}* and *{patient, postmenopausal, fear of cancer}* would be allowed to participate in a potential query, since there are outgoing recursive links from both concept terms *{premenopausal, postmenopausal}* leading to the concept term *{fear of cancer}.* Note that *assembled properties* and/or *values* are also enabled in potential queries.

Considerations for the set of operational terms I_F: Given that all domain and operational terms underlie a further classification based on their meaning, an assignment of operational terms to operands (domain terms) is defined in terms of associations (links) holding among classes C_T of operational and domain terms which belong to the corresponding layers at the top of Figure 4.

For example, given that an operational term is classified as *Univariate Function,* it can be applied only to single properties within a current query context, since there is an association holding between the class *{Single Property}* as subclass of *{Property}* and the class of operational terms *{Univariate Function}.* An operational term which is member of the class *Bivariate Function* can be applied only to terms classified as *{Concept}* or *{Complex Property}.* Moreover, operational terms classified as *{Comparison Operator}* can be assigned only to terms classified as *{Domain Values}.*

In general, potential consideration of a particular operation/operator to be assigned to a particular domain alphabet term within a current query state is a matter of classification of both the operational term(s) and the domain alphabet term, as well as the association holding among them. In other words, these associations stand for the sense of *which operations/operators could be applied over which domain alphabet terms.*

Since the nature of operational terms is application domain independent, classification for these terms as well as the holding associations between operational and domain terms classification layers are constant.

An example is given, in the following, of holding associations at the classification layers of operational and domain terms. Within the domain terms classification layer

{Domain Values} → {Atomic Value, Interval, Complex Value}
{Domain Values} → {Qualitative Value, Quantitative Value}
{Property} → {Single Property, Complex Property}
{Property} → {Categorical Variable, Numerical Variable},
whereas within the operational terms classification layer
{Interrelationship Function} → {Bivariate Function, Multivariate Function}
and between classification layers of operational and domain terms
{Domain Values} → {Comparison Operators}
{Single Property} → {Univariate Function}
{Complex Property} → {Interrelationship Function}
{Entity} → {Interrelationship Function}

Representation of Constraints

Since *meaning* is also expressed in terms of constraints, which make the connection of two objects through a link in the knowledge space relative to predefined conditions, *preconditions* can also be assigned to the objects of the knowledge space.

Representation of constraints has been investigated within the context of *Constraints Satisfaction Problems* (CSP; Tsang, 1993) for solving a number of problems arising in the areas of operational research, policy making, optimization, etc. Constraints are mainly expressed by a set of variables and values connected with arithmetic and/or comparison operators. The problem solution is expressed in terms of {*variable, value*} pairs, which satisfy all constraints.

In the context of **MDDQL**, constraints are currently expressed in terms of information objects as members of the set of alphabet terms. Constraints are conceived as *preconditions* the satisfaction of which determines their final consideration for inclusion in a potential subsequent query state. Since one of the major focuses of the Meaning Driven Querying Methodology has been the construction of a query in terms of a *semantically consistent* description of instances, we must exclude *semantic inconsistencies* in terms of *mutually exclusive* alphabet terms. Therefore, the consideration of a particular concept, property, value or operational term is made relative to the *query context*, i.e., the set of current query terms already considered by the end user and constituting the current query state.

We mainly distinguish between two classes of preconditions: a) those that apply to domain terms and b) those that apply to operational terms. Note that preconditions are expressed in terms of *term unique identifiers*–natural language independent expressions–but for the sake of understanding, the terms expressed in English have been used.

Consider, for instance, the query addressing the instances described as *premenopausal patients with asymptomatic leiomyomata and estimated uterine weight less than 300 grams and a growth of 20% within 12 months* should be rejected as semantically meaningless, since the value term *20% within 12 months* is not semantically consistent with the value *less than 300 grams* assigned to the property *estimated uterine weight*.

Definition: A *domain term* including precondition p_t is conceived as a set of literals constituting a propositional formula in conjunctive $p_t = (p_1 \wedge p_2 \wedge ... \wedge p_m)$ (*CNF*) or disjunctive $p_t = (p_1 \vee p_2 \vee ... \vee p_m)$ normal form (*DNF*), where the propositional variable p_i is defined as $p_i = \cup(t_n)$, $t_n \in T, i \in 1, 2, ..., m$. It also holds that $p_t \in P_T$, where P_T is the set of all specified domain terms including preconditions over a particular domain of alphabet terms.

Definition: Given a query state q_s, the *truth functionality semantics* of p_t is defined in terms of the mapping of $p_i \rightarrow true$, due to the appearance of $p_i = \cup(t_n)$ within the current query context q_s, otherwise $p_i \rightarrow false$.

On the other hand, since operational terms are also conceived as information objects, they become part of the information space considered so far. Therefore, constraints can be assigned to operational terms similarly to those constraints posed for domain terms. However, the definition of preconditions slightly departs from the definition of preconditions defined over domain terms.

Since assignment of operations to operands is done at the level of classification structures of terms rather than instances of terms, a precondition over operational terms is also defined in terms of classification roles. For example, if the affected domain term is classified as property and as *categorical variable*, then univariate functions such as *calculation of average* must be excluded from suggestion. Similarly, if the affected operand is a term classified as domain value and is currently assigned to a property classified as *categori-*

cal variable, then comparison operators such as *greater than* or *less than* must be excluded from suggestion.

A more detailed description of the formalisms underlying the specification of preconditions and/or circumstances is given in Kapetanios (2000).

THE INFERENCE ENGINE

Given the knowledge space as structured and described in the previous section, the inference engine works on the basis of various algorithms depending on the complexity of queries to be posed to the system. Its main task is to infer and, subsequently, suggest terms to the end users which are not only semantically consistent with the current state of the query but also with all those semantic elements which might contribute to a thorough understanding of particular terms of the vocabulary in a particular natural language.

It is out of the scope of this chapter to cover all the aspects of the inferences made by the inference engine of **MDDQL**. However, there is an example of an implemented algorithm in the following which illustrates the inferences made during the system-guided construction of a simple query (no operations are considered). The token <**SELECT**> refers to the choices made by the end user. Given that we are dealing with *patients' profiles*, the entry point to the knowledge space will be always the node representing the term *Patient*.

Algorithm:

Start query construction

Suggest set of natural languages

e.g., {*English, German*}

<**SELECT**> a natural language in which terms should appear

e.g., {*English*}

Suggest further entity terms to start with

e.g., {*Coronary angiography, Revascularization*}

<**SELECT**> an entity term

e.g., {*Coronary angiography*}

Move to the initial query state q_0

e.g., q_0 = {*Patient, Coronary angiography*}

Set initial query state q_0 to be the current state q_c

Repeat

<**SELECT**> a term in current query state $t_S \in q_C$

while there are outgoing edges from $t \in q_c$ to some nodes (terms) with $t \in I_D$ **do**

get connected node

if connected node is not already in q_c **then**

if there is a set of preconditions assigned to node **then**
    if set of preconditions is satisfied
        add node to list of inferred nodes (terms) T_I
    else ignore node
  else
    add node to list of suggested nodes
 else ignore node
end of while
Suggest inferred terms T_I to end user for selection as information objects (section 5.1)
e.g., {*atypical angina, chronic stable angina, acute myocardial infarction,*
...}
with t_s = *Coronary angiography*
<SELECT> term(s) from T_I
e.g., {*atypical angina*}
Add selected term(s) to the current query state q_c by assigning them to t_s
e.g., {*Patient, Coronary angiography, atypical angina*}
Move to q_j which is a subsequent query state of q_c
Make q_j the current query state $q_c \equiv q_j$
until final query state is reached
e.g., {*Patient, Coronary angiography, atypical angina, gender, female, stress test, positive*}

Following this example, the terms {*gender, stress test*} belong to a set of suggested (inferred) terms T_I when t_s = {*atypical angina*}, $t_s \in q_c$ and $q_c \equiv$ {*Patient, Coronary angiography, atypical angina*}, whereas the term {*female*} belongs to the set of suggested (inferred) terms when t_s = {*gender*}, $t_s \in q_c$ and $q_c \equiv$ {*Patient, Coronary angiography, atypical angina, gender*}. Similarly, the term {*positive*} belongs to the set of suggested (inferred) terms when t_s = {*stress test*}, $t_s \in q_c$ and $q_c \equiv$ {*Patient, Coronary angiography, atypical angina, gender, female, stress test*}.

QUERY LANGUAGE EXPRESSIVENESS

Up to now, we had a close look at the interactive mode of query construction due to the represented knowledge concerning the nature of domain application and operational terms. The contructed query, however, always resembles a semantic query graph (tree) having the nodes carrying most of the semantic information which is needed in order to transform the **MDDQL** query into a data repository specific one such as SQL.

Therefore, the set of all query trees that can be constructed out of the knowledge space of terms (section 5) represents the set of all possible queries that can be submitted to the system. In the following, the structure of MDDQL query trees is given in terms of **BNF** notation. "*" means any number of occurences, "{}" means optionally, "|" means alternatively. Paths are defined as binary, i.e., connecting two adjacent nodes and having as common root a concept node.

```
<query> ::= {<analytical operation>} (<path> {<path>})*
<node> ::= <concept> | <relationship> | <property> | <value>
<path> ::= <concept><concept> | <concept><relationship> |
<relationship><concept> | <relationship><relationship> |
<concept><property> | <relationship><property> |
<property><property> | <property><value>
<concept> ::= (<tui><word><reference>{<logical operator>})
<relationship> ::= (<tui><word><reference>)
<property> ::= (<tui><word><reference>{<logical operator>}
{<univariate function>})
<value> ::= (<tui><word><reference>)*{<comparison opera-
    tor>}
<comparison operator> ::= > | < | <> | ≥ | ≤ | between | like
<logical operator> ::= AND | OR | NOT
<reference> ::= {<repository>}<implementation symbol>
<analytical operation> ::= <bivariate function> | <multivari-
ate function>
```

For example, the semantic query tree which corrensponds to the query as expressed in free text with *Appropriateness of coronary angiography for patient with atypical angina having gender female, stress test positive and age under 50* is depicted in Figure 5. For the construction of this query, only application domain terms which belong to the terminology base coronary angiography have been used.

Given the query construction mode as described previously, the **MDDQL** query trees are considered to be semantically meaningful, since they have been constructed due to the represented knowledge (semantics) characterizing the terminology base for both application domain and operational terms. Given the structural properties of the query trees as well as the information carried by the nodes, it is possible to map each **MDDQL** query to queries formulated in an instances description query language which rely on the operations of the relational algebra (union, intersection, difference, projection, selection, join) such as SQL. A formal proof of the **MDDQL** equivalence to the relational algebra outlies the scope of this chapter.

On the other hand, it is also possible to map the **MDDQL** query tree structure to XML data structures and, therefore, to XML-based query

Figure 5: An example MDDQL query tree

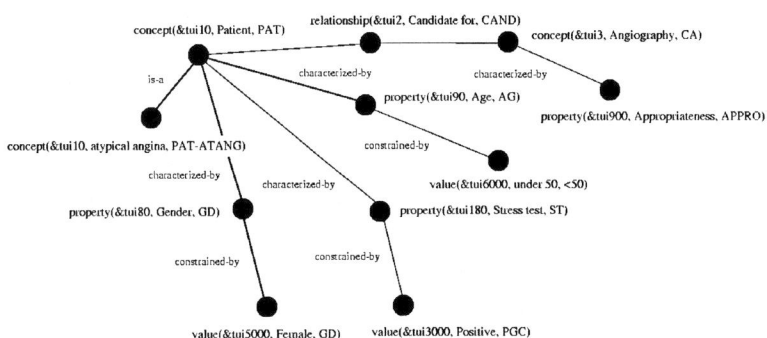

languages. This is due to the fact that the XML data model is conceived as a tree with the difference that edges carry the most information as related to XML data structures (Abiteboul, Buneman & Suciu, 2000), in contrast with the semantic query trees of **MDDQL** where the nodes (vertices) carry the most information. A mapping of **MDDQL** query trees to XPath-based queries is straightforward.

FUTURE TRENDS

The Semantically Advanced Query Answering System with the Meaning Driven Data Query Language (**MDDQL**) as a query language is conceived as a knowledge providing and validating system, where the key issue is the role of semantics in both query formulation and answer generation. However, such a system poses additional challenges to the components of

- hypothesis (expressed by a classification rule) validation/verification based on real data,
- enhancement of the query execution engine by data mining algorithms and/ or statistical procedures,
- application discourse meaning and semantics modeling tool which meets the requirements of the meaning representation model presented in this chapter,
- enhancement of answer presentation mechanism by semantically advanced visualization techniques,
- query transformation and execution engine when more than one data repository is involved,

which are currently tested.

In particular, querying in terms of hypothesis (classification rule) validation against evidence-based data would enhance the already existing querying mechanism of classification rules. This is based on time/spatial constraints as well as on probability theory. Furthermore, statistical operations and/or data mining algorithms will be integrated within the query alphabet and will be considered by the query execution engine.

Another important issue is the maintenance of knowledge space and inference algorithms, since changes of domain semantics result from the state-of-the-art medical practices. However, changes would affect only the data/knowledge structures, since the inference algorithms operate on structural and semantic properties of data which are represented separately (terminology base). In other words, there is a clear distinction between data/knowledge structures and algorithms. At the moment, there are no maintenance tools implemented which enable updates of the knowledge space. This is due to the fact that the knowledge representation model makes use of semantic constraints which are not supported by currently available tools for ontological or knowledge engineering.

The currently available visual query interfaces are being adapted in order to provide the facility of starting the query construction process with any term out of the set of domain alphabet terms. Since the major philosophy of the query construction process with MDDQL is to incrementally refine the query context in terms of the major concepts (themes) we are focusing on, starting with an arbitrary term means that the system will try first to clarify the context of the term through the suggestions as made by the inference engine. For instance, if the entry term would have been *stress test*, then the system should suggest any relevant concepts that are brought into association with this term.

Tightly coupled with this approach is the consideration of terms where more than one source is taken into account. The problem of semantic heterogeneity as stated by homonyms (same elements mean different things) and synonyms (similar things are represented by the same elements) found in different repositories is conceived as a knowledge-based suggestion of related terms according to query context as formed by the set of all terms appearing in the participating repositories.

CONCLUSION

A semantically advanced query language for medical knowledge and decision support has been presented in this chapter which enables a system-guided construction (formulation) of meaningful queries based on *meaning* and/or *semantics* of terms as they appear in medical application discourses. Representation of meaning and/or semantics is achieved by the structure of a knowledge space which is consulted by an inference engine which drives the query construction process. Both the knowledge space and the inference engine underlie the specification of **MDDQL** as the Meaning Driven Data Query Language at the core of the system. The language meets the requirements of

- no learning of a particular query or natural language based syntax for querying a data repository,
- no understanding of the underlying database schema in terms of adequate interpretations of data constructs such as relations, classes, collections,
- embedding natural language based interpretation of the meaning of acronyms used as attribute and/or values,
- awareness of the context which relates to the data as expressed in terms of measurement units, explanations or definitions, constraints, etc.

The system has been and is currently in use in various cardiological and gynecological hospitals and clinics in Switzerland. In particular, the system has been used for evaluation purposes of medical interventions concerning coronary angiography and revascularization by the *Cardiocentro Ticino* (**CCT**) in conjunction with the *Ospedale Civico di Lugano* as well as by the *Triemlispital Zurich*. The system is currently in use by the gynecological clinics *Bruderholz, Schaffhausen, Unispital Zurich*, etc.

REFERENCES

Abiteboul, S., Buneman, P. and Suciu, D. (2000). *Data on the Web: From Relations to Semistructured Data and XML*. Morgan Kaufmann Publishers.

Abiteboul, S., Hull, R. and Vianu, V. (1995). *Foundations of databases*. Addison-Wesley Pub. Company.

Abiteboul, S., Quass, D., McHugh, J., Widom, J. and Wiener, J. (1997). The lorel query language for semistructured data. *International Journal on Digital Libraries*, 1(1), 68-88.

Angelaccio, M., Catarci, T. and Santucci, G. (1990). QBD*: A graphical query language with recursion. *IEEE Transactions on Software Engineering*, 16(10), 1150-1163.

Badre, A., Catarci, T., Massari, A. and Santucci, G. (1996). Comparative ease of use of a diagrammatic vs. an iconic query language. In *Interfaces to Databases. Electronic Series Workshop in Computing,* 1-14, Springer.

Banhart, F. and Klaeren, H. (1995). A graphical query generator for clinical research databases. *Methods of Information in Medicine*, 34(4), 328-

339.

Batini, C., Catarci, T., Costabile, M. and Levialdi, S. (1991). Visual query systems: a taxonomy. In *Proc. of the 2nd IFIP 2.6 working conference on visual databases,* 153-168, Budapest: North Holland.

Berztiss, A. T. (1993). The Query Language VIZLA. *IEEE Transanctions on Knowledge and Data Engineering,* 5(5), 813-825.

Cardiff, J., Catarci, T. and Santucci, G. (1997). Semantic query processing in the VENUS environment. *International Journal of Cooperative Information Systems,* 6(2), 151-192.

Catarci, T., Costabile, M., Levialdi, S. and Batini, C. (1997). Visual query systems for databases: A survey. *Journal of Visual Languages and Computing,* 8(2), 215-260.

Catarci, T. and Santucci, G. (1988). Query by diagram: A graphic query system. In *Proc. of the 7th conference on the entity-relationship approach* (291-308). Roma, Italy: North Holland.

Catarci, T., Santucci, G. and Angelaccio, M. (1993). Fundamental graphical primitives for visual query languages. *Information Systems,* 18(2), 75-98.

Catell, R. and Barry, D. K. (1997). *The Object Database Standard: ODMG 2.0.* Morgan Kaufmann Publishers, Inc.

Ceri, S., Comai, S., Damiani, E. and Fraternali, P. (1999). XML-GL: A Graphical Language for Querying and Restructuring WWW Data. In *Proc. of the 8th Inter. World Wide Web Conference,* WWW8, Toronto, Canada.

Chan, H. (1997). Visual query languages for entity relationship model databases. *Journal of Network and Computer Applications,* 20(2), 203-221.

Chan, H., Siau, K. and Wei, K. (1998). The effect of data model, system and task characteristics on user query performance-An empirical study. *DATA BASE FOR ADVANCES IN Information Systems,* 29(1), 31-49.

Chavda, M. and Wood, P. (1997). Towards an ODMG-Compliant Visual Object Query Language. In Jarke, M., Carey, M., Dittrich, K., Lochovsky, F., Loucopoulos, P. and Jeusfeld, M. (Ed.), *Proc. of 23rd international conference on very large data bases,* 456-465.

Chu, W. W., Merzbacher, M. A. and Berkovich, L. (1993). The design and implementation of CoBase. In *ACM SIGMOD* 93 (517-522). Washington D.C.

Chu, W. W., Yang, H., Chiang, K., Minock, M., Chow, G. and Larson, C. (1996). CoBase: A scalable and extensible cooperative information system. *Journal of Intelligent Information Systems*.

Cinque, L., Levialdi, S. and Ferloni, F. (1991). An expert visual query system. *Journal of Visual Languages and Computing*, 2, 101-113.

Clark, G. J. and Wu, C. T. (1994). DFQL-data-flow query language for relational databases. *Information and Management*, 27(1), 1-15.

Clark, J. (1999). *XML Path Language (XPATH)*. Retrieved on the World Wide Web: http://www.w3.org/TR/xpath.

Dennebouy, Y., Anderson, M., Auddino, A., Dupont, Y., Fontana, E., Gentile, M. and Spaccapietra, S. (1995). SUPER: Visual interfaces for object + relationship data models. *Journal of Visual Languages and Computing*, 6, 74-99.

Doan, D. K., Paton, N. W., Kilgour, A. C. and Qaimari, G., et al. (1995). Multi-paradigm query interface to an object-oriented database. *Interacting with Computers*, 7, 25-47.

Elmarsi, R. and Wiederhold, G. (1981). GORDAS: A formal high-level query language for the entity-relationship model. In *Proceedings of the 2nd International Conference on Entity-Relationship Approach*, 49-72, Washington DC: USA.

Epstein, R. (1991). The TableTalk query language. *Journal of Visual Languages and Computing*, 2, 115-141.

Florescu, D., Raschid, L. and Valduriez, P. (1996). A methodology for query reformulation in CIS using semantic knowledge. *International Journal of Cooperative Information Systems*, 5(4), 431-467.

Gil, I., Gray, P. M. and Kemp, G. J. (1999). A visual interface and navigator for the P/FDM object database. In Paton, N. and Griffiths, T. (Eds.), *Proceedings of the 1st International Workshop on User Interfaces to Data Intensive Systems, (UIDIS 99)*, 54-63. Edinburgh, Scotland: IEEE Computer Society Press.

Groette, I. and Nilsson, E. (1988). SICON: An icon presentation module for an E-R database. In *Proceedings of the 7th International Conference on Entity-Relationship Approach*, 271-289. Roma, Italy.

Groff, J. R. and Weinberg, P. N. (1999). *SQL: The Complete Reference*. Osborne/McGraw-Hill.

Group, W. X. W. (1998). The query language position paper of the XSL working group. In *Proceedings of the Query Languages Workshop*. Cambridge, MA.

Haw, D., Goble, C. and Rector, A. (1994). GUIDANCE: Making it easy for the user to be an expert. In Sawyer, P. (Ed.), *Proceedings of the 2nd International Workshop on Interfaces to Database Systems*, 19-43,

Springer Verlag.

Houben, G. and Paredaens, J. (1989). A graphical interface formalism: specifying nested relational databases. In Kunji, T. (Ed.), *Visual Data-Base Systems*, North-Holland.

Hripcsak, G., Allen, B., Cimino, J. J. and Lee, R. (1996). Access to data: Comparing AccessMed with query by review. *Journal of the American Medical Informatics Association*, 3(4), 288-299.

Juristo, N. and Moreno, A. M. (2000). Introductory paper: Reflections on conceptual modeling. *Data and Knowledge Engineering* (33), 103-117.

Kapetanios, E. (2000). *A Meaning Driven Querying Methodology*, Unpublished Doctoral thesis, ETH-Zurich, Switzerland.

Kapetanios, E., Norrie, M. and Fuhrer-Stakic, D. (2000). MDDQL: A visual query language for meta-data driven querying. In *5th IFIP International Working Conference on Visual Database Systems*. Fukuoka, Japan: Kluwer Academic Publisher.

Kifer, M., Kim, W. and Sagiv, Y. (1992). Querying object-oriented databases. *Proc. ACM SIGMOD Symposium on the Management of Data*, 393-402.

King, R. and Melville, S. (1984). SKI: A semantic knowledgeable interface. In *Proceedings of the 10th VLDB,* (30-37). Singapore.

Kobler, A. and Norrie, M. (1999). OMS Java: Lessons learned from building a multi-tier object management framework. In *Workshop on Java and Databases: Persistence Options, OOPSLA 1999*, Denver, USA.

Kromrey, H. (1994). *Empirische Sozialforschung*. Opladen: Leske + Budrich. (6. revidierte Auflage).

Li, W. and Shim, J. (1998). Facilitating complex Web queries through visual user interfaces and query relaxation. *Computer Networks and ISDN Systems*, 30(1-7), 149-159.

Liepins, P., Curran, K., Renshaw, C. and Maisey, M. (1998). A browser based image bank, useful tool or expensive toy? *Medical Informatics*, 23(3), 199-206.

Loukopoulos, P. and Karakostas, V. (1995). *Systems Requirements Engineering*. McGraw-Hill.

Maier, D. (1998). Database desiderata for an XML query language. In *Proceedings of the Query Languages Workshop*. Cambridge, Mass.

Massari, A., Pavani, S., Saladini, L. and Chrysanthis, P. (1995). QBI: Query by icons. In *Proceedings of the ACM SIGMOD Conference on Management of Data,* 477. San Jose, USA: ACM Press.

Meoevoli, L., Rafanelli, M. and Ricci, F. L. (1994). An interface for the direct manipulation of statistical-data. *Journal of Visual Languages and Computing*, 5(2), 175-202.

Merz, U. and King, R. (1994). DIRECT-A query facility for multiple databases. *ACM Transactions on Information Systems*, 12(4), 339-359.

Murray, N., Goble, C. and Paton, N. (1998). A framework for describing visual interfaces to databases. *Journal of Visual Languages and Computing*, 9(4), 429-456.

Murray, N., Paton, N. and Goble, C. (1998). Kaleidoquery: A visual query language for object databases. In *Proceedings of Advanced Visual Interfaces* (25-27), Italy.

Mylopoulos, J., Borgida, A., Jarke, M. and Koubarakis, M. (1990). TELOS: Representing knowledge about information systems. *ACM Transactions on Office Information Systems*, 8(4).

Norrie, M. (1993). An extended entity-relationship approach to data management in object-oriented systems. In *12th International Conference on Entity-Relationship Approach*, 390-401. Springer Verlag.

Norrie, M. (1995). Distinguishing typing and classification in object data models. In *Information Modeling and Knowledge Bases* (VI). IOS.

Ozsoyoglou, G. and Wang, H. (1993). Example-based graphical database query languages. *Computer*, 26(5), 25-38.

Papantonakis, A. and King, P. (1995). Syntax and semantics of GQL, A graphical query language. *Journal of Visual Languages and Computing*, 6, 3-25.

Robie, J., Lapp, J. and Schach, D. (1998). XML query language (XQL). In *Proceedings of the Query Languages Workshop*, Cambridge, MA.

Schilling, J., Faisst, K., Kapetanios, E., Wyss, P., Norrie, M. and Gutzwiller, F. (2000). Appropriateness and necessity research on the Internet: Using a second opinion system. *Journal of Methods of Information in Medicine*.

Shirota, Y., Shirai, Y. and Kunii, T. (1989). Sophisticated form-oriented database interface for non-programmers. In Kunji, T. (Ed.), *Visual Database Systems*, 127-155. North-Holland.

Siau, K. L., Chan, H. C. and Tan, K. P. (1992). Visual knowledge query language. *IEICE Transactions on Information and Systems*, E75D(5), 697-703.

Taira, R., Johnson, D., Bhushan, V., Rivera, M., Wong, C., Huang, L., Aberle, D., Greaves, M. and Goldin, J. (1996). A concept-based retrieval system for thoracic radiology. *Journal of Digital Imaging*, 9(1), 25-36.

Tsang, E. (1993). *Foundations of Constraint Satisfaction*. Academic Press, London.

Wegner, L. (1989). ESCHER: Interactive visual handling of complex objects in the extended NF2 database model. In Kunji, T. (Ed.), *Visual Database Systems,* 277-297. North-Holland.

Wieringa, R. (1995). *Requirements Engineering: Frameworks for Understanding*. Chichester: Wiley.

Yu, C. T. and Meng, W. (1998). *Principles of Database Query Processing*. Morgan Kaufmann Publishers Inc.

Zhang, G., Chu, W. W., Meng, F. and Kong, G. (1999). Query formulation from high-level concepts for relational databases. In Paton, N and Griffiths, T. (Eds.), *Proceedings of the International Workshop on User Interfaces to Data Intensive Systems, uidis 99,* 64-74. Edinburgh, Scotland: IEEE Computer Society Press.

Zloof, M (1977). Query-by-example: A database language. *IBM Systems Journal*, 16(4), 324-343.

Chapter VII

Towards a Secure Web-Based Healthcare Application[1]

Konstantin Knorr
University of Zurich, Switzerland

Susanne Röhrig
SWISSiT Informationstechnik AG, Switzerland

INTRODUCTION

In healthcare a lot of data are generated that in turn will have to be accessed from several departments of a hospital. The information kept within the information system of a hospital includes sensitive personal data that reveal the most intimate aspects of an individual's life. Therefore, it is extremely important to regard data protection laws, privacy regulations, and other security requirements. When designing information systems for healthcare purposes, it is an imperative to implement appropriate access control mechanisms and other safeguards. Furthermore, a tendency to use the Internet as a communications media can be observed. As the Internet is an insecure transmission media, the security requirements that must be met by the overall system are high.

During the project MobiMed (Privacy and Efficiency of Mobile Medical Systems; further information about the project can be found at http://www.ifi.unizh.ch/ikm/MobiMed/), a prototype was developed to show the feasibility of the implementation of security mechanisms required in a Web-based healthcare application.

This chapter describes the specific security requirements in healthcare environments, focusing on the additional security demands resulting from the use of the Internet as a communications media.

RELATED WORK

The technical committee 251 of the European Committee for Standardization is responsible for medical informatics. Working group 6 focuses on standardization in healthcare concerning security, privacy, quality and safety. The safety of systems is defined as "the expectation that systems do not, under defined conditions, enter a state that could cause human death or injury." We will not discuss "safety" in this chapter, but focus on "security" instead, whose definition is the subject of the next section *General Security Objectives.*

Rindfleisch (1997) gives a general introduction of privacy aspects and security measures in healthcare. Anderson (1996b) introduces a security policy model for clinical information systems which makes use of nine principles. The policy is compared to traditional policies from the banking and military sectors. Traditionally, security requirements in healthcare environments are high. Especially since the rise of global networks a lot of attention has been paid to this topic. Baker and Masys (1999), Rind (1997), and Blobel (1997) present distributed systems and networks in healthcare regarding security aspects focussing on confidentiality and privacy.

The prototype that will be introduced in this chapter is based on a clinical process. Healthcare processes and their support through information and communication technology have recently received much attention. Scheer, Chen and Zimmermann (1996) discuss the process management in a hospital. Knorr, Calzo, Röhrig and Teufel (1999) describe the modeling and redesign of a clinical process. Kuhn, Reichert and Dadam (1995), Yousfi, Geib and Beuscart (1994), Reichert, Schultheiss and Dadam (1997), and Bricon-Souf, Renard and Beuscart (1998) introduce, discuss and analyze workflow management systems and their impact on healthcare.

Nevertheless the above publications focus only on specific security objectives. Unlike these approaches we take a general and structured outlook on the problem and take into account the four major security objectives: confidentiality, integrity, availability, and accountability. These objectives and their special meaning in healthcare systems are regarded, specifically for data transmission and storage. A prototype shows possible security measures and the feasibility of their implementation.

GENERAL SECURITY OBJECTIVES

Generally, when talking about data security–in healthcare as well as in other areas–the three objectives confidentiality, integrity, and availability are identified. This division dates back to the early 1980s and was described in National Bureau of Standards (1980). In this section we will describe these security objectives and illustrate their specific meaning in a healthcare environment.

Confidentiality: Confidentiality is defined as the state that exists when data is held in confidence and is protected from unauthorized disclosure. If the content of a communication is disclosed to an unauthorized person or even if the fact that a communication between two persons took place is made public, this is considered a loss of confidentiality.

In healthcare many sensitive data are processed, leading to high requirements on the confidentiality of data. Traditionally, the need for confidentiality stems from national data protection legislation and the professional secrecy of medical staff. Most data protection laws also state that data may only be used for the purpose they were collected for, e.g., data collected during a medical treatment may not be used for marketing purposes.

Integrity: Integrity is the state that exists when data has not been tampered with or has been computed correctly from source data and not been exposed to accidental or malicious alteration or destruction. Erroneous input and fictitious additions are also considered violations of data integrity. The demand for integrity is also included in some data protection laws (e.g., Switzerland).

The modification (whether with malicious intent or due to a program failure) of data stored within a healthcare application may lead to a mistreatment of the respective patient which could be hazardous to his health. In a case published in Der Spiegel (1994) an attack on a hospital information system in Liverpool was described. The attacker changed the data of prescriptions. A patient was saved only because a nurse rechecked the prescription and did not administer the prescribed medication, realizing that it was lethally toxic.

Availability: When all required services and data can be obtained within an acceptable period of time, the system is called available. Data concerning a patient have to be available at any time, to prevent loss of time in case of emergencies, since a treatment without knowing certain medical data could harm the patient's health. Anderson (1996a) describes what may happen if the availability of a healthcare system is not assured, citing the collapse of the London Ambulance Service in October and November 1992. Due to the overload and collapse of a new computerized dispatching system, London was left with partial or no ambulance cover for longer periods, which is believed to have led to a loss of about 20 lives.

Recently, more security objectives have been identified. The most important one is accountability.

Accountability: If the accountability of a system is guaranteed, every participant of a communication can be sure that his partner is exactly the one he or she pretends to be. This allows holding users responsible for their actions. In healthcare accountability mechanisms like that are necessary because it is important to know who performed a certain service at a certain time. Today, manual signatures of the responsible person ensure accountability.

ADDITIONAL REQUIREMENTS FOR DISTRIBUTED APPLICATIONS

When transferring medical data over the Internet, additional security requirements arise. This mainly touches the following two areas:

Secure data transfer: If data are collected at different sites the need for data transfer arises when the data are processed at a central location. The data transmitted might be subject to certain risks–especially if public networks (e.g., the Internet) are used as transportation media.

Secure data storage: If the sites of data collection and of data processing are different, it has to be ensured that the storage on the other side of the communication is authorized. This depends on data protection laws as well as the professional secrecy in healthcare, which is embodied in the penal legislation of many countries. Also the other security objectives have to be regarded.

Moreover, the computers taking part in data transfer have to meet the same security requirements as any other computer used in healthcare. However, these have been described thoroughly (e.g., Anderson, 1996b) and are not topic of this chapter.

In the following subsections the aspects "secure data transfer" and "secure data storage" are regarded more closely.

Secure Data Transfer

Additional security requirements when transferring medical data over the Internet–organized according to the security objectives identified in the section *General Security Objectives* are as follows.

Confidentiality: When data is exchanged over the Internet, it is generally accessible to everybody who has access to the network. The main threats to confidentiality for data on the Internet are presented in Damm et al. (1999).

Integrity: Data transferred over the Internet not only are read easily but are also manipulated as easily. Even though the data are protected against transmission errors on the lower layers of the TCP/IP (Transmission Control Protocol/Internet Protocol) protocol family, intentional damage can not be prevented as an attacker may easily recompute check sums to forge data (Damm et al., 1999).

Availability: To ensure a certain availability the Internet was built redundantly. However, the current implementations of TCP/IP allow attackers to disturb the operation of computers or parts of the network, e.g., through denial-of-service (DoS) attacks like SYN-flooding (Damm et al., 1999). A well-known example of how the availability of a major part of the Internet may be compromised is the *Morris Worm Incident* of November 1988, where a self-replicating program used design flaws of BSD-derived versions of UNIX and disrupted normal Internet connectivity for days (Spafford, 1988).

Accountability: In a distributed system especially, problems identifying the originator of a message might occur. On the Internet the techniques to produce a wrong IP address of the sender's computer are well known (so-called IP spoofing). Moreover, the address could be changed any time during the transmission.

Secure Data Storage

More security requirements arise when collecting data at one site and storing them centrally at another.

Confidentiality: When medical data are collected at one site and stored at another, the regulations of the data protection legislation have to be followed.

In Switzerland the use of medical data for research purposes is allowed only in specific cases that are listed in Der Eidgenössische Datenschutzbeauftragte (1997). In contrast, the security policy of the British National Health Service states that a patient in treatment implicitly agrees to the use of his or her data for research purposes (Anderson, 1996b).

The use of anonymized data, however, is allowed under most laws; anonymized meaning that the patient cannot be reidentified from the data stored. We therefore conclude that–though they have different focuses–the national data protection legislation has to be followed, i.e., as few data as possible have to be collected about a patient and the access to them has to be as restricted as possible.

Integrity: The data stored at the receiver's side has to be kept in a state of integrity, and the user sending these data has to trust that they are not changed. An attacker might use a known system vulnerability to gain access to the system, where he could modify data.

Additional integrity may be achieved by the use of plausibility controls before storing.

Availability: The availability of data and services at the storing side has to be kept up at any time, as the users in the hospitals must have access in case any problems occur. Recent incidents where the availability of commercial Internet servers was disturbed are the distributed denial-of-service attacks against Internet merchants in February 2000. Among others the Internet bookstore amazon.com and the Internet auctioneer ebay.com could not be accessed during several hours (Spiegel On-line, 2000).

Accountability: The server where data is stored has to prove to the users that it is the one it pretends to be. Otherwise an attacker might run a denial-of-service attack, take the regular server out of order, and pretend to be this server–consequently receiving data not intended for it.

GENERAL SECURITY MEASURES

To counter the risks described above, appropriate security measures have to be implemented.

Protection of confidentiality: Access control mechanisms, encryption techniques, or anonymization are generally used to protect confidentiality. Access control mechanisms grant or restrict access to data or applications on either the application or operating system layer. Encryption techniques are used to prevent unauthorized persons from reading data not intended for them. Anonymization means that any reference to the concerned person is removed from the data.

Protection of integrity: The integrity of data (e.g., messages) can be guaranteed by the use of cryptographic check sums (so-called message authentication codes). A hash value of the message is calculated and appended. The receiver again calculates the hash value of the message. By comparison with the sent value, transmission errors can be detected. In connection with public-key encryption such hash values may be used as digital signatures.

Protection of availability: Within a distributed system the availability of the computers and the communications media must be guaranteed. Measures to protect the availability of a system are either of organizational and technical nature or concern the surrounding infrastructure. Organizational and technical measures include:

- appropriate configuration of any hard- and software used on the system,
- regular use of security software (such as scanners or intrusion detection systems),
- protection of the whole system from outside attack by using a well-configured firewall system,
- rules concerning passwords,
- backup frequencies and the secure preservation of backup media, and
- documented procedures what to do in case of security violations.

Measures concerning the infrastructure are, e.g., the safe placement of server systems and cables as well as measures for fire protection.

In addition, one principle design concept of the Internet was redundancy, i.e., the data packets may find multiple ways from sender to recipient, so that one damaged routing element might not disturb the whole communication chain. Thus, redundancy increases the availability of the system.

Protection of accountability: Measures to guard the accountability within a system are log files–where all activity is recorded–and digital signatures together with appropriate public-key infrastructures and certificates.

Since confidentiality is traditionally the most important security aspect of medical data, the main focus of the security measures, described in more detail in the section *Implementation*, lies on the protection of the confidentiality of the data stored and transmitted.

The following paragraphs depict access rights, encryption, and certification, which are also the main security mechanisms implemented in the MobiMed prototype.

Access Control Mechanisms

A well-known concept to grant or deny access rights is the use of role-based access control together with an access control matrix. Such a matrix is built in a way that the objects to be accessed (i.e., the data or the applications of the system) are listed on one axis and the roles on the other axis. The access rights are organized as the matrix entries. Each user is assigned one or more roles; he or she is granted access only if one of his or her roles has access to the requested data.

This approach can be enhanced if not only the user's role but also the state of the process of work is considered when granting or denying access rights. Hence, the access is only granted to persons who–at a given time–need this data to accomplish their tasks. The access control becomes dependent on the context. Thomas and Sandhu (1993) first described this concept. Holbein (1996) suggests granting access rights according to the context of a business process. A necessary prerequisite is the analysis and modeling of the underlying processes. This modeling is done using a workflow management system.[2] In comparison to an access control system based purely on role-based access control matrices, this scheme offers additional protection against insiders trying to misuse stored data.

A system using workflow states as the basis to grant access rights was described by Nitsche, Holbein, Morger, and Teufel (1996) using the ActionWorkflow System (Medina-Mora, Winograd, Flores & Flores, 1992).

Encryption

To protect data from being read by unauthorized persons, encryption techniques can be used. Two different classes of encryption are distinguished: symmetric and asymmetric algorithms.

When using a symmetric algorithm both parties of a communication have to agree beforehand on one common key, which is used for both encrypting and decrypting the data. Common symmetric encryption schemes are DES (Data Encryption Standard) and IDEA (International Data Encryption Algorithm). An advantage of symmetric encryption is that it is rather fast; a disadvantage, however, is the difficult key management as keys have to be agreed upon by each pair of potential communication partners and distributed in a secure manner.

An asymmetric algorithm demands that each user owns his or her own pair of keys: a private key, which is kept secret and known only by its user, and a public key, which is published and can be accessed by everyone who wants to communicate with the key's owner. Those schemes are called public-key algorithms. A well-known public-key algorithm is RSA (named after its inventors: Ron Rivest, Adi Shamir, and Leonard Adleman). When a message is encrypted with one of the keys, it can only be decrypted using the other key. This feature can be used to implement confidentiality as well as accountability: If a message has to be kept secret, it will be encrypted using the recipient's public key, so that only he can decrypt it, as no-one else knows his private key. If it has to be proven that a certain person sent a message, she will encrypt it using her private key; everybody who decrypts it will use her public key, thereby ensuring that it was she who wrote the message. A disadvantage of public-key algorithms is that they are rather slow to compute, whereas the key management is easier compared to symmetric algorithms.

To combine the advantages of symmetric and asymmetric algorithms, so-called hybrid techniques are used. An asymmetric scheme is used to negotiate a symmetric session key, which is later used to encrypt the data being sent. This combines the simple key management of asymmetric techniques and the fast encryption of the data being sent with symmetric techniques.

A thorough description of cryptographic algorithms can be found in Schneier (1994).

Certification

To establish trust when using public-key crypto-systems, it is necessary that the users are given a guarantee that a public key really belongs to the person the key is issued to. Certificates issued by a trusted third party like a CA (Certification Authority) offer a solution.

Such an authority signs the users' keys with its own private key, thereby certifying that the information about the user is correct and that the public key really belongs to the respective user. Another user can now check the certificate by verifying the signature, thus ensuring that the key really belongs to its user. Of course, the CA must be a trustworthy entity that certifies keys only after the user has identified himself. The CA's own public key may in turn be signed by another CA, thereby creating either a hierarchy of CAs or a cross-certification structure. An infrastructure where trust is established and users can access and check other users' keys is called a PKI (Public Key Infrastructure).

Another task of a CA is the management of cryptographic keys. More information about certification and public key infrastructures can be found in Branchaud (1997). Fumy and Landrock (1993) describe the principles of key management.

The structure of certificates issued by CAs is standardized. One important standard is the X.509 authentication framework to support the X.500 directory services (Branchaud, 1997). An X.509 certificate comprises, among other data, a certificate serial number, the identifier of the signature algorithm, the certificate's validity period, the user's name, and the user's public key information. All this is signed with the CA's private key.

APPLICATION SCENARIO

Clinical Research Cycle

The application developed during our project supports a clinical study–a so-called *clinical research cycle* (CRC). In the scope of this study a new medication is tested on volunteers within a hospital (Raven, 1991). This test is carried out in four steps:

1. During the first part, the test person is checked for eligibility, whether he is eligible for the study. For this end, certain questions concerning the patient's anamnesis, inclusion and exclusion criteria are examined.
2. Consequently, a pre-examination is carried out, recording a laboratory analysis of the patient's blood and other tests before the patient is treated with the medication to be tested.

3.	Subsequently, the medication is administered and the same tests as in step 2 are carried out.

4.	During a post-examination a few weeks after the medication administration, the same tests are carried out for a third time.

This practice allows analyzing the impact of the medication on the patient's health. The test results are then sent for analysis to the pharmaceutical company who commissioned the test. Traditionally, the CRC is carried out with paper-based questionnaires. This procedure is error-prone and tedious because large amounts of paper are used and the data inserted in the questionnaires has to be reentered manually into an electronic system at the pharmaceutical company. In our prototype the questionnaires are electronic–to be more specific they have been implemented as HTML (Hypertext Markup Language) forms with text areas, radio buttons, and select boxes as input options. A sample screenshot of an input mask is shown in Figure 1.

In our scenario the data are collected at several hospitals and are sent to a central WWW and database server to be stored there. No data of the CRC are kept at the hospitals.

This process is a good example for distributed processing as the test may be carried out not only in one but in several hospitals, e.g., to find an appropriate number of test persons. When processing data in a distributed manner, special data protection and security requirements arise that are topic of this chapter.

The questionnaires of the CRC contain highly sensitive data concerning the patient, like his history of drug abuse and mental condition. Therefore, it is very important to regard data protection regulations.

A patient volunteers to participate in the CRC. Before any examination he explicitly consents that his data may be stored. In this consent the patient acknowledges that his anonymized medical data can be used for the given purpose. Only data raised during the CRC are stored in the prototype's database.

The User Interface of the Prototype

The prototype developed during our research project supports the execution of the CRC described in the last subsection. This subsection briefly discusses the user interface and the operational framework of the prototype.

The user interface is completely based on HTML so that standard Web browsers can be used. To use the prototype, a user has to enter the URL (Uniform Resource Locator) of the prototype's Web site in his browser.

When a user logs onto the system (user name and password are required), a menu offers the following alternatives:

1.	Browse the description of the underlying CRC. Both the process and the logical and temporal order of the questionnaires are explained.

Figure 1: Screenshot of the prototype

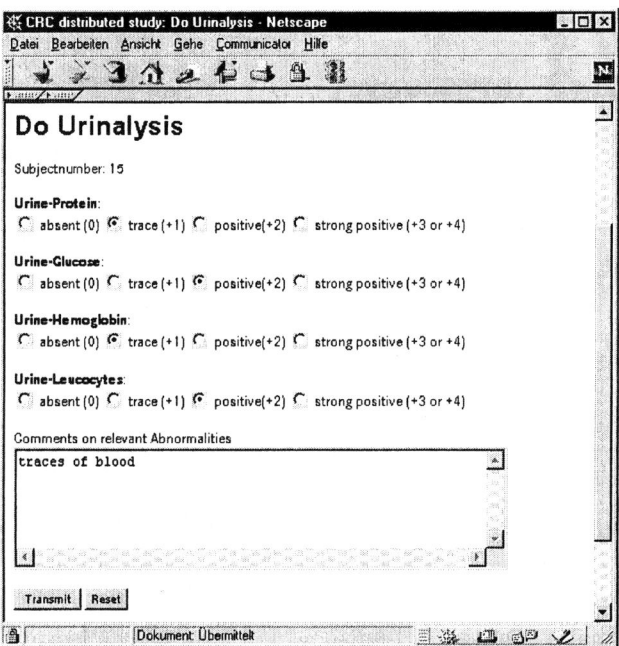

2. Show the work list depending on the user's role (e.g., "doctor" or "nurse"). Following a hyperlink in the work list automatically opens the corresponding questionnaire. Figure 1 shows an example. Each entry in the work list represents a patient (identified through his initials and date of birth) and the possible steps of the examination.
3. Browse data that already exists (from other questionnaires in the study) if access rights are granted.

IMPLEMENTATION

In the last sections security was examined in a very thorough but abstract way. Practical considerations will be the topic of this section. The implementation of the security measures is shown according to the prototype developed in our research project.

The architecture of the prototype is shown in Figure 2. It is a typical client-server application. The client may be any Web browser capable of SSL (Secure Sockets Layer). The server side consists of a HTTP (Hypertext Transfer Protocol) server, several Perl (Practical Extraction and Report Language) scripts (Wall, 1996), and a DBMS (Database Management System). The database stores the medical data associated with the test of the new medication and the workflow data (more precisely the model of the underlying process) needed for the context-dependent access control. The Perl scripts act like a "glue layer" between the WWW and the database server. The following example illustrates the way the different components of the prototype work together. A typical client request is treated in six steps:

1.   The client requests some data from the medical database. This request is done via a special HTML form. The data packets exchanged between browser and server are SSL-encrypted.

2.   The WWW server receives the request and forwards the relevant entries in the HTML form to a Perl script. The whole communication between WWW server and the Perl script makes use of CGI (Common Gateway Interface) technology.

3.   The Perl script generates an SQL (Structured Query Language) statement out of the different entries in the HTML form, establishes an ODBC (Open Database Connectivity) connection to the database server, and transfers the SQL statement to the DBMS.

4.   The database server executes the SQL statement and returns the result containing the requested data items to the Perl script.

5.   The Perl script "translates" the result of the SQL statements to HTML and forwards the HTML output to the WWW server.

6.   The WWW server hands the dynamically generated HTML page back to the browser. The data are SSL-encrypted, again.

The server side of the application is running on a PC using the operating system Windows NT 4.0. The WWW server is the Internet Information Server 4.0 and the DBMS is SQL Server 6.5. Perl version 5 is used.

Figure 2: Architecture of the MobiMed prototype

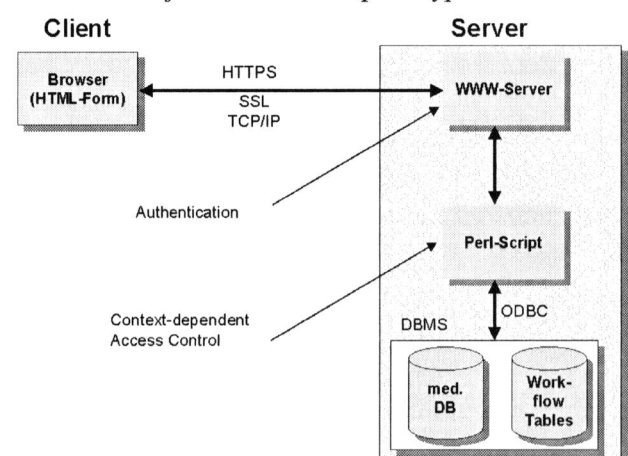

After this rather technical description of the prototype its main security mechanisms are considered in detail.

How these mechanisms correspond to the four security objectives–confidentiality, integrity, availability, and accountability–is shown in Table 1. A bullet (•) indicates that the security mechanism implements the security objective.

Workflow-Based Access Rights

In our prototype access rights are granted according to the state of the underlying CRC. Therefore, a model of the process has to be created. Such a model is called a workflow specification. The main issues of such a specification, according to Jablonski, Böhm, and Schulze (1997), are:

- activities,
- data items,
- participants,
- control flow, and
- data flow.

Note that the workflow specification has to be defined before the workflow is executed. During the execution many instances of the workflow model are generated. In our prototype one workflow instance is created for each patient.

In the CRC the activities correspond to the questionnaires to be completed (Figure 1 shows the activity "Do Urinalysis"). The data items are the information contained in the questionnaires; the participants are the three user roles "doctor," "nurse," and "monitor of the pharmaceutical company." The control flow determines in which sequence the activities must be executed, whereas the data flow states which already existing data items are needed to carry out a new activity.

The state of a workflow describes which activities have been performed and which activities are ready for execution. For better handling, the states of the CRC have been numbered in our prototype.

The context-dependent access control is done using a database table which consists of the four entries *user role*, *state of the workflow*, *name of the questionnaire*, and *kind of access right*. For instance, (nurses, 3, urinalysis, write) is a possible quadruple in the table, i.e., a nurse has write permission on the database table which contains the laboratory data of the urine of a patient, if the workflow of the patient is in phase three of the CRC.

We use this strict means of access control to ensure the confidentiality in our system, thus not only keeping outsiders from viewing the data but also restricting insiders to using the data only when necessary, i.e., the access to the data is as restricted as possible.

Table 1: Security objectives implemented in the prototype

		security objectives			
		confidentiality	integrity	availability	accountability
security mechanisms	workflow-based access rights	•			
	SSL	•	•		•
	authentication	•			•
	log files				•
	further security measures	•	•	•	

SSL

The CRC can be executed in different hospitals. The Internet–a public network–is used to communicate with the central database of the pharmaceutical company. As stated before, the Internet is a highly insecure network. Specific security considerations have to be made. To protect the integrity and authenticity of data packets on the Internet, a special security protocol called SSL is also used.

SSL provides data encryption, server authentication, message integrity, and optional client authentication for a TCP/IP connection and is located above TCP/IP and below HTTP on the protocol stack. A hybrid encryption scheme is used; typically the keys have a length of 512 bits (asymmetric) and 40 bits (symmetric). To prove the authenticity of the server (possibly of the client, too), a certificate from a trusted third party must be obtained. For more information about SSL see Internet Draft (1996).

For our prototype a certificate by the Swiss CA *SwissKey* (http://www.swisskey.ch) was obtained, which proves that the Web server is run by a trusted entity known to the CA.

We use SSL not only to keep the communication confidential but also to prove that the server is the same it claims to be–thus enforcing accountability within the system. The use of SSL also enhances the integrity of the data transmitted through message authentication codes.

Authentication

Only legitimate users are allowed to communicate with the WWW server of the prototype. To prove their legitimization users have to provide a valid log-in name and password every time they start a session with the server. Both username and password are transmitted in an encrypted way–using the SSL connection–so that no attacker can easily gain a user-name and a password by intercepting the communication. Authentication is a measure to increase confidentiality and accountability of our application.

Users should be encouraged to choose a password which cannot be guessed easily, because an attacker gaining knowledge of a valid user password can act on behalf of that user (Morris & Thompson, 1979).

Log Files

All user activity dealing with sensitive medical data is logged by the application. Unsuccessful or attempted data access is logged, too. The log file contains the following entries:
- the name and role of the user,
- the time and date of access,
- the performed activity,
- the IP address and domain name (if available) of the client,
- the patient identification number,
- the phase to the patient's workflow at the access, and
- the accessed database table.

The log file can be used to provide evidence of what data was requested, accessed, or inserted by a user at a specific time. This is used to increase the system's accountability. Furthermore log files can be used to do load management if several servers are involved, to do recovery if an unexpected error occurred, and to fulfill certain legal documentation requirements.

Further Security Measures

The data to be protected in our prototype are stored in a central database. This database is located at a secure site distinct from the hospitals where the test of the new medication is done. There is a physical separation of the locations where data are raised and stored. Hence, it is very difficult for a user to tamper with data he has already transferred to the database.

The security of a server system as well as the applications on the server can be tested through special security scanners. One prominent example is the scanner distributed by *Internet Security Systems* (http://www.iss.net). To exclude as many known system vulnerabilities as possible, such a security scan has been performed on the computer hosting both the WWW and the database server–increasing the availability of the system.

Furthermore, when submitting data to the server, they are checked for completeness, thus enhancing the integrity of the system.

When using our prototype in practice, it is very important to implement a system to back up data to minimize the risk of data loss. Most database systems provide means to support databack up. If the WWW server has a high workload, backup or standby servers may be necessary, too.

The use of CGI scripts may provide opportunities to attack the system. Special "script scanner software"–like Whisker–is available to check all scripts on a server for these vulnerabilities (http://www.wiretrip.net/rfp/). More ideas to enhance the prototype's security without the use of CGI-scripts are described by Ultes-Nitsche and Teufel (2000).

Another important aspect is anonymization: In a clinical trial, no personal patient data at all is submitted to the WWW server (and consecutively to the pharmaceutical company). The patient's personal information is kept at the hospital. On the server only a unique identifier is used. The medical data cannot be matched to the person's name and demographic data. Therefore, all data protection requirements are met.

CONCLUSION

In this chapter we have concentrated on security requirements of Web-based healthcare applications. The main focus has been laid on a systematical presentation of the security objectives: confidentiality, integrity, availability, and accountability.

These objectives have been defined, and their implications on data transfer and data storage have been shown. The implementation of the security requirements was illustrated with our prototype, an Internet-based application supporting the data management associated with the testing of a new medication in a hospital. The prototype's main security features are SSL encryption of the communication between client (WWW browser) and server (WWW and database server), password authentication at the WWW server, the use of log-files, and the so-called context-dependent access control which yields additional protection for the sensitive medical data in the database.

The confidentiality on the server is ensured by a sophisticated access control mechanism; encryption protects the confidentiality during the transfer of the data.

Data integrity is provided for by plausibility checks on the server side and by message authentication codes offered by SSL during the transmission. Physical protection of the server and the use of special security software to scan the system for vulnerabilities enhance the availability of the system. The use of SSL and certificates, login procedures and the use of a log file ensure the system's accountability. Table 1 gives an overview of the security measures implemented in the prototype and the security objectives aimed at.

To sum up, the prototype implements different security mechanisms to counter the four major security objectives in structured and layered security architecture.

FUTURE RESEARCH

Future research will focus on the following topics:
1. The presented security requirements and the aspects of data transfer and data storage will be presented and extended in a structured matrix, yielding a systematic model to organize and measure the security of distributed applications (Knorr & Röhrig, 2000).
2. The prototype could furthermore be extended with digital signatures to permit the so-called "four eyes principle," which is very important in healthcare scenarios. To introduce digital signatures in the prototype, a public-key infrastructure has to be established. Also, elaborate authentication mechanisms like smart cards or biometric devices for authentication purposes will be considered, assuming that the necessary browser interfaces will be provided.
3. A third research topic will be the legal requirements of healthcare applications in different European countries. A typical example is the legal status of digital signatures, which varies from country to country.

ACKNOWLEDGMENTS

The authors thank Prof. Kurt Bauknecht and the Information and Communication Management group of the University of Zurich for their help and support. Additionally, we would like to thank Dr. Rolf Grütter for his trust and cooperation over the last years. Konstantin Knorr's work was partially sponsored by the Swiss National Science Foundation under grant SPP-ICS. 5003-045359.

ENDNOTES

1 A preliminary version of this chapter appeared in (Röhrig & Knorr, 2000).
2 An overview of Workflow Management Systems is given in (Georgakopoulos, Hornick & Sheth, 1995).

REFERENCES

Anderson, R. J. (1996a). Information technology in medical practice: Safety and privacy lessons from the United Kingdom. *Australian Medical Journal.* Retrieved on the World Wide Web: http://www.cl.cam.ac.uk/users/rja14/austmedjour/austmedjour.html.

Anderson, R. J. (1996b). A security policy model for clinical information systems. *Proceedings of the 1996 IEEE Symposium on Security and Privacy*, 30-43.

Baker, D. B. and Masys, D. R. (1999). PCASSO: A design for secure communication of personal health information via the Internet. *International Journal of Medical Informatics*, 54(2), 97-104.

Blobel, B. (1997). Security requirements and solutions in distributed electronic health records. *Proceedings of the IFIP/Sec 1997*, 377-390.

Branchaud, M. (1997). *A Survey of Public-Key Infrastructures.* Unpublished master's thesis, McGill University, Montreal.

Bricon-Souf, N., Renard, J. M. and Beuscart, R. (1998). Dynamic workflow model for complex activity in intensive care unit. *Proceedings of the Ninth World Congress on Medical Informatics.* Amsterdam: IOS Press, 227-231.

Damm, D., Kirsch, P., Schlienger, T., Teufel, S., Weidner, H. and Zurfluh, U. (1999). Rapid secure development–Ein verfahren zur definition eines Internet-sicherheitskonzeptes. *Projektbericht SINUS.* Institut für Informatik, Universität Zürich, Technical Report 99.01.

Der Eidgenössische Datenschutzbeauftragte. (1997). Leitfaden für die Bearbeitung von Personendaten im medizinischen Bereich. Bearbeitung von Personendaten durch private Personen und Bundesorgane. Bern.

Der Spiegel. (1994). (9), 243.

Fumy, W. and Landrock, P. (1993). Principles of key management, *IEEE Journal on Selected Areas in Communication*, 11(5).

Georgakopoulos, D., Hornick, M. and Sheth, A. (1995). An overview of workflow management: From process modeling to workflow automation infrastructure. *Distributed and Parallel Databases*, 3, 119-153.

Holbein, R. (1996). *Secure Information Exchange in Organizations–An Approach for Solving the Information Misuse Problem.* Unpublished doctoral thesis, University of Zurich.

Internet Draft. (1996). *The SSL Protocol,* Version 3.0. Retrieved on the World Wide Web: http://home.netscape.com/eng/ssl3/draft302.txt.

Jablonski, S., Böhm, M. and Schulze, W. (1997). *Workflow-Management.* Heidelberg: Dpunkt Verlag für digitale Technologie.

Knorr, K., Calzo, P., Röhrig, R. and Teufel, S. (1999). Prozeßmodellierung im Krankenhaus. Electronic business engineering/4. *Internationale Tagung Wirtschaftsinformatik.* Physica-Verlag, 487-504.

Knorr, K. and Röhrig, S. (2000). Security of electronic business applications-structure and quantification. *Proceedings of the 1st International Conference on Electronic Commerce and Web Technologies (EC-Web 2000),* Greenwich, UK, 25-37.

Kuhn, K., Reichert, M. and Dadam, P. (1995). Unterstützung der klinischen Kooperation durch Workflow-Management-Systeme-Anforderungen, Probleme, Perspektiven. Tagungsband der 40. Jahrestagung der GMDS, 437-441.

Medina-Mora, R., Winograd, T., Flores, R. and Flores, F. (1992). The action workflow approach to workflow management technology. *Proceedings of the 92 CSCW Conference.*

Morris, R. and Thompson, K. L. (1979). Password security: A case history, *Communications of the ACM,* 22(11), 594-597.

National Bureau of Standards. (1980). Guidelines for security of computer application. *Federal Information Processing Standards Publication 73,* Department of Commerce.

Nitsche, U., Holbein, R., Morger, O. and Teufel, S. (1996). Realization of a context-depended access control mechanism on a commercial plattform. *Proceedings of the IFIPSec '96,* Chapman & Hall.

Raven, A. (1991). *Consider it Pure Joy–An Introduction to Clinical Trials.* Cambridge: Raven Publication.

Reichert, M., Schultheiss, B. and Dadam, P. (1997). Erfahrungen bei der Entwicklung vorgangsorientierter, klinischer Anwendungssysteme auf Basis prozessorientierter Workflow-Technologie. Tagungsband zur GMDS-Jahrestagung 1997, München: MMV Medizin Verlag.

Rind, D. M. (1997). Maintaining the Confidentiality of Medical Records Shared over the Internet and the World Wide Web. *Annals of Internal Medicine,* 127, 138-141.

Rindfleisch, Th. C. (1997). Privacy, information technology and healthcare. *Communications of the ACM,* 40(8), 93-100.

Röhrig, S. and Knorr, K. (2000). Towards a secure Web-based healthcare application. *Proceedings of the 8th European Conference on Information Systems (ECIS 2000–A Cyberspace Odyssey)*, Vienna, 1323-1330.

Scheer, A. W., Chen, R. and Zimmermann, V. (1996). *Prozessmanagement im Krankenhaus. Krankenhausmanagement.* Wiesbaden: Gabler-Verlag. 75-96.

Schneier, B. (1994). *Applied Cryptography.* New York: Wiley.

Spafford, E. H. (1988). *The Internet Worm Program: An Analysis, CS-TR-833*, Department of Computer Science, Purdue University, West Lafayette.

Spiegel On-line (2000). *Attackiert: Das Web im Fadenkreuz.* Retrieved on the World Wide Web: http://www.spiegel.de/netzwelt/technologie/ 0,1518,63447,00.html.

Thomas, R. K. and Sandhu, R. S. (1993). Towards a task-based paradigm for flexible and adaptable access control in distributed applications. *Proceedings of the 1992-1993 ACM SIGSAC New Security Paradigms Workshop*, Little Compton, RI, 138-142.

Ultes-Nitsche, U. and Teufel, S. (2000). Secure access to medical data over the Internet. *Proceedings of the 8th European Conference on Information Systems (ECIS 2000–A Cyberspace Odyssey)*, Vienna, 1331-1337.

Wall, L. (1996). *Programming Perl.* Cambridge, MA: O'Reilly.

Yousfi, F., Geib, J. M. and Beuscart, R. (1994). A new architecture for supporting group work in the field of healthcare. *Proceedings of the Third Workshop on Enabling Technologies: Infrastructure for Collaborative Enterprises.*

Chapter VIII

Secure Internet Access to Medical Data

Ulrich Ultes-Nitsche
University of Southampton, United Kingdom

Stephanie Teufel
University of Fribourg, Switzerland

INTRODUCTION

In Holbein et al. (1997) and previous papers (Holbein & Teufel, 1995; Holbein et al., 1996), the concept of a context-dependent access control has been introduced and discussed exhaustively. A prototype implementation of the concept is described in Nitsche et al. (1998). The prototype implementation is for local use only and would reveal many security holes if used over an open network: The dynamic link library (DLL) that handles the access control, for instance, would be publicly accessible. In Nitsche et al. (1998), by spying out the DLL code, one could obtain information about the database's administrator log-in procedure, possibly leaving the entire database unprotected. However, using technology different from that presented in Nitsche et al. (1998) allows one to come up with a secure distributed solution to context-dependent access control over the Internet.

In context-dependent access control, information about the state of a business process is combined with general knowledge about a user to grant or revoke access to sensitive data. The basic concepts of such an access control scheme are understood very well (Holbein, 1996). However, when developing a concrete system one faces several problems: Existing access control mechanisms of the target platform have to be adapted to support context dependency, missing features have to be realized in some indirect way, etc.

It is the aim of the current chapter to present an implementation concept for context-dependent access control on the Internet. Even though applied to the specific application area of clinical trials, the underlying concepts are general and support all applications for which context-dependent access control is suitable. The chapter summarizes a part of the Swiss National Science Foundation funded project *MobiMed*[1] (Fischer et al., 1995). The system we are going to describe is PC-based (Windows NT) and implemented as a Java servlet accessing an MS SQL Server database. The servlet extends the functionality of Java Web server, which makes it accessible from the Internet. Context information, which is used for checking the authority of an access request, is delivered by an Action Technology workflow system.

This chapter emphasizes the novel *technical* solution to accessing medical data over the Internet under a context-dependent access control policy. So we concentrate very much on describing the impact that the use of Java servlets has. More aspects of context-dependent access controls can be found in Röhrig and Knorr (2000). After a short introduction into context-dependent access control, we present the basic implementation concepts. To give a detailed description of the implementation of context-dependent access controls, we give an overview of the components that ensure security. Finally, we discuss the appropriateness of the presented solution.

CONTEXT-DEPENDENT ACCESS CONTROL

Role-based security approaches fit well the hierarchically structured setting of a hospital (Ting et al., 1991, 1992). Each level in the hierarchy can be mapped to a so-called organizational role that a person at this hierarchical level plays in the hospital (e.g., medical personnel, care personnel, etc.). After an analysis of each role's demands on obtaining particular data to perform work, access rights are assigned to each role. The access rights determine which records in the database that contains information about the patients (patient records) may be read or written by a person playing a particular role. When logging in, a user of the system identifies her- or himself by using a chipcard and a PIN, and then a role is assigned to the person according to user/role lists, determining her or his access rights. A different way to handle role assignment is to store role information on the chipcard itself in ciphered form, which then is read during log-in or, alternatively, whenever data records are accessed.

Simple role-based access control mechanisms have the advantage that they can be implemented rather easily but the drawback of certain inflexibilities. A more sophisticated access control technique refining the role-based approach takes into account an access request's particular point in time, i.e., the question: "Is it reasonable that a person playing a particular role *needs* access to certain data at the current state of a healthcare process?"[2] Obviously, it is not necessary to have access to all data about a particular patient all the time. The question of "What does one *need to know*?" gave the considered principle its name: the *need-to-know* principle. Moreover, what does one *need to know right now* (at the time of an access request) is the context-dependent access control scheme that we consider in this chapter. A possible realization of the need-to-know principle can be achieved by combining user role information with state information of a workflow ("Is it resonable to grant a person, playing a particular role, access to particular data in the given workflow state?"). A workflow system, which can be used to determine context information, is the Action Workflow approach (Medino-Mora et al., 1992).

THE ACCESS-CONTROL SYSTEM IN PRINCIPLE

As mentioned above, context-dependent access control combines user information with process state information to compute access rights. Since the database system that we consider—MS SQL Server—uses group-based access control, we have to build context dependency around group-based

Figure 1: The group concept of MS SQL Server

access granted to group *care personnel* ↓	access granted to group *medical personnel* ↓
Department/Room	**Type of Examination**
GastroEnt / 12 b	Gastroscopy
Medicine I / A323	Gastroscopy
GastroEnt / 12 b	Colonoscopy
Medicine II / B411	Gastroscopy

access controls. In group-based access control, a list of group names is assigned to a table or table entry of a database. A group is a name to a list of user names. If a user requests access to a table entry, the group-based access control tries to match the user name with a group assigned to the requested data for the type of access request. If the user name can be matched, access is granted.

In Figure 1, the group *care personnel* has access to table *Department/ Room* and group *medical personnel* has access to table *Type of Examination*. Usually, medical personnel have access to all the data that care personnel can access. By making group *medical personnel* a subgroup of group *care personnel*, *medical personnel* inherits all access rights of *care personnel*. This is sketched in Figure 2.

Subsequently, we describe how the group concept of MS SQL Server can be used to realize context dependency. The first constraint we have to consider is that group assignments in a given hospital should not be changed: The access rights to existing tables for which no context-dependent access control is established should be left unchanged. On the other hand, for new tables in

Figure 2: Inheriting access rights

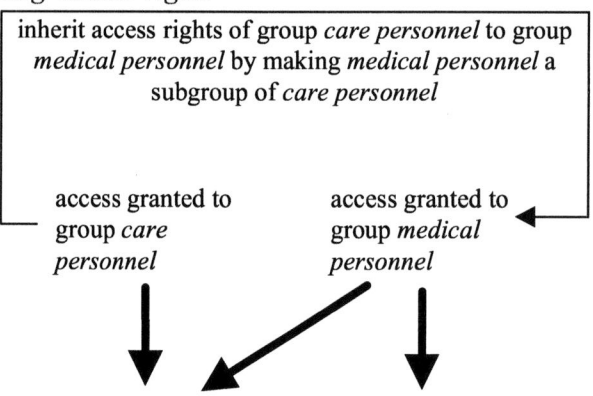

Department/Room	Type of Examination
GastroEnt / 12 b	Gastroscopy
Medicine I / A323	Gastroscopy
GastroEnt / 12 b	Colonoscopy
Medicine II / B411	Gastroscopy

the database, context-dependent access control shall be established, implying that for none of the existing groups access is granted a priori to these new tables.

Access rights in the context-dependent scheme are only given temporarily to users for a single access. We achieve this by creating a not yet existing group, called here *MobiMedDBAccessGranted*. No user and no group is assigned to *MobiMedDBAccessGranted* initially and permanently. In order to change the group membership of a user temporarily, we can use the SQL stored procedure `sp_change_group`. After checking the *need to know* of a user to perform an access (Holbein et al., 1997), he or she receives temporary membership in the group *MobiMedDBAccessGranted*, the access is performed, and the user is reset to her/his previous group.

As mentioned earlier, the context information that is used in the evaluation of the authority of an access request is taken from a business process model. In such a model, information about the history of events to a particular instantiation of the process is stored in a process state. In the context of a hospital, the state comprises information about treatment of a patient, and to each patient a single instantiation of the business process is created. The process ID of the instantiation is used as a unique key to all database entries related to the patient. As usual, access control tables are generated in the context-dependent framework containing more information than usual. The context-dependent access-control table contains quadruples *stateID*, *groupID*, *table entry*, *access type*. These entries have the meaning that in state *stateID*, a user belonging to group *groupID* may access database entry *table entry* by an access of type *access type*. The *access type* indicates whether full access, read-only access, or no access at all will be granted.

Based on the concept sketched in this section, we will develop subsequently a realization of the Internet context-dependent access control that is secure and feasible. To implement it, we use Java servlets, securing the communication using SSL.

JAVA SERVLETS

The past year has seen the rise of server-side Java applications, known as Java servlets. Servlets are used to add increased functionality to Java-enabled servers in the form of small, pluggable extensions. When used in extending Web servers, servlets provide a powerful and efficient replacement for CGI and offer many significant advantages (Sun Microsystems, 1998). These advantages include:

Portability: Java servlets are protocol and platform independent and as such are highly portable across platforms and between servers. The servlets must conform to the well-defined Java servlet API[3], which is already widely supported by many Web servers.

Performance: Java servlets have a more efficient life cycle than either CGI or FastCGI scripts. Unlike CGI scripts, servlets do not create a new process for each incoming request. Instead, servlets are handled as separate threads within the server. At initialization, a single object instance of the servlet is created that is generally persistent and resides in the server's memory. This persistence reduces the object creation overhead. There are significant performance improvements over CGI scripts in that there is no need to spawn a new process or invoke an interpreter (Hunter, 1998). The number of users able to use the system is also increased because fewer server resources are used for each user request.

Security: The Java language and Java servlets have improved security over traditional CGI scripts both at the language level and at the architecture level.

Language Safety: As a language Java is type safe and handles all data types in their native format. With CGI scripts most values are treated and handled as strings, which can leave the system vulnerable. For example, by putting certain character sequences in a string and passing it to a Perl script, the interpreter can be tricked into executing arbitrary and malicious commands on the server.

Java has built-in bounds checking on data types such as arrays and strings. This prevents potential hackers from crashing the program, or even the server, by overfilling buffers. For example, this can occur with CGI scripts written in C where user input is written into a character buffer of a predetermined size. If the number of input characters is larger than the size of the buffer, it causes a buffer overflow and the program will crash. This is commonly known as stack smashing.

Java has also eliminated pointers and has an automatic garbage collection mechanism, which reduces the problems associated with memory leaks and floating pointers. The absence of pointers removes the threat of attacks on the system where accesses and modifications are made to areas of server memory not belonging to the service process.

Finally, Java has a sophisticated exception handling mechanism, so unexpected data values will not cause the program to misbehave and crash the server. Instead an exception is generated which is handled and the program usually terminates neatly with a run-time error (Garfinkel & Spafford, 1997).

Security Architecture: Java servlets have been designed with Internet security issues in mind and mechanisms for controlling the environment in which the servlet will run have been provided.

CGI scripts generally have fairly free access to the server's resources and badly written scripts can be a security risk. CGI scripts can compromise the security of a server by either leaking information about the host system that can be used in an attack or by executing commands using untrusted or unchecked user arguments. Java significantly reduces these problems by providing a mechanism to restrict and monitor servlet activity. This is known as the servlet sandbox. The servlet sandbox provides a controlled environment in which the servlet can run and uses a security manager to monitor servlet activity and prevent unauthorized operations. There are four modes of operation that include trusted servlets, where the servlet has full access to the server resources, and untrusted servlets, which have limited access to the system.

JDK 1.2 is introducing an extension to its security manager, the access controller. The idea behind the access controller is to allow more fine-grained control over the resources a servlet can access. For example, instead of allowing a servlet to have write permission to all files in the system, write permission can be granted for only the files required by the servlet for execution (Hunter, 1998).

However, Java-based servers are still vulnerable to denial-of-service attacks, where the system is bombarded with requests in order to overload the server resources. This approach invokes so many servlet instances that all the server resources are allocated. This can impact all the services supported by the server. However, the effects of this can be reduced by specifying an upper limit on the number of threads that can be run concurrently on the server. If all the threads are allocated, that particular service can no longer be accessed, but because the server still has resources left to allocate, the rest of the services are still available.

THE SECURE SOCKETS LAYER PROTOCOL

The Secure Sockets Layer Protocol (SSL) is designed to establish transport layer security with respect to the TCP/IP protocol stack. Version 3 was published as an Internet draft document (Freier et al., 1996) by the IETF (Internet Engineering Task Force). We introduce SSL briefly along the lines of Stallings (1998) and motivate its usage for the MobiMed prototype.

The Protocol Stack: The transport layer part of SSL, the SSL Record Protocol, sits on top of TCP in the Internet protocol stack. It is accessed by an upper layer consisting of the Hypertext Transfer Protocol (HTTP) and different parts contributing to SSL: SSL Handshake Protocol, SSL Change Cipher Spec Protocol, and the SSL Alert Protocol, used to set up, negotiate, and change particular security settings used by the SSL Record Protocol. Schematically, the SSL architecture is presented in Figure 3.

Different Security Features of SSL: SSL allows for different security

Figure 3: SSL within the Internet protocol stack (Stallings, 1998)

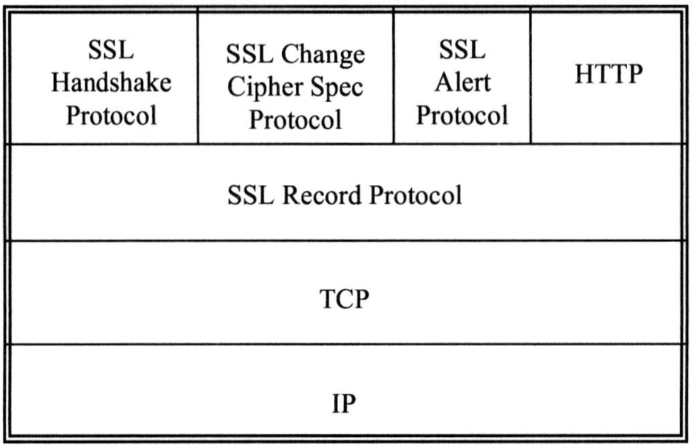

SSL Handshake Protocol	SSL Change Cipher Spec Protocol	SSL Alert Protocol	HTTP
SSL Record Protocol			
TCP			
IP			

features being chosen. First of all, different encryption algorithms can be used to produce ciphertexts and authentication messages. For authentication, different hash algorithms can be negotiated. SSL can also use X509.v3 peer certification (Garfinkel & Spafford, 1997). Using a session identifier, active states of SSL are identified, where a state consists of a number of keys involved in the session, both on the server and on the client side, and sequence numbers to count the messages exchanged. By using these different parameters, SSL sets up a session configuration that then allows for ensuring integrity, confidentiality, and authentication depending on the setup parameters.

Use of SSL in MobiMed: Unlike other concepts that secure connections or even only data packages, SSL includes the concept of a secure session, determined by the parameters mentioned in the subsection above. It is this session concept that makes it appealing for being used in MobiMed. The secure session lasts as long as a user is logged in the system. Since communication with the user will be based on HTML documents sent to and received from the client side by using HTTP, the use of SSL will be transparent to the client.

REALIZING THE ACCESS-CONTROL SYSTEM

To perform the SQL stored procedure `sp_change_group`, one needs to have the access rights of the SQL Server's administrator. These access rights could not be granted if the context-dependent access control itself were implemented as an SQL stored procedure due to SQL Server restrictions. Hence we decided to implement it in a Java servlet. The servlet offers the only way to access the *MobiMedDB* database and can be accessed from

Figure 4: The system interfaces

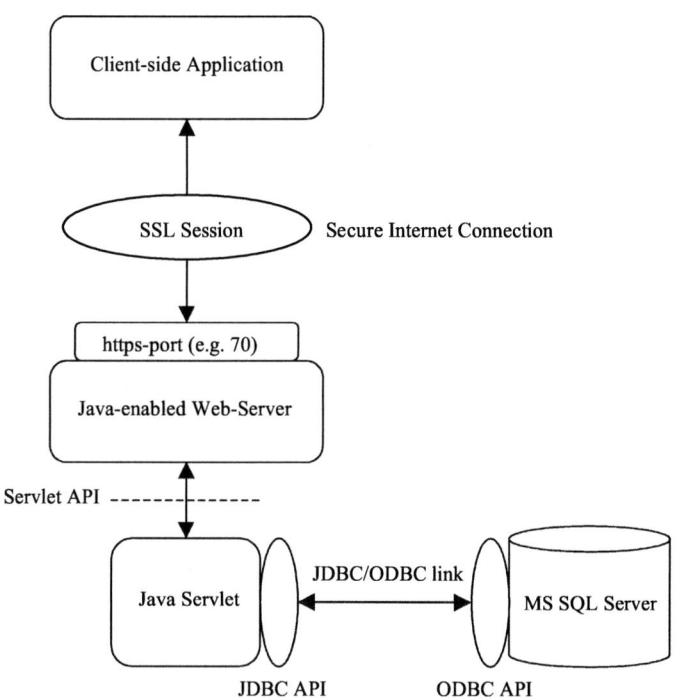

any application on the Internet. It must know the administrator log-in procedure of the SQL Server. By putting this procedure into the private part of the servlet class, it is securely protected, since this part, by no means, can be sent to the client side. We overcome by this the problem of spying out administrator log-in information, a problem that existed in the local-version prototype of this system (Nitsche et al., 1998; Ultes-Nitsche & Teufel, 2000).

Java servlets support the *Java DataBase Connectivity* (JDBC) API to access databases that support the JDBC API. SQL Server supports the *Open DataBase Connectivity* (ODBC) API that can be linked to the JDBC API. The interfaces within the resulting system are presented in Figure 4.

The client-side application (either HTML or a Java applet) talks to the servlet via the Java Web server. The connection between the client and the Web server is secured using SSL. The servlet offers to the client application the standard servlet API. This is implemented by extending the *HttpServlet* class. The servlet API offers to handle HTTP *get* and *post* requests (the methods of the *HttpServlet* class are *doGet* and *doPost*). The parameters sent to the servlet are in our case SQL queries that are interpreted by the servlet. As presented in Figure 4, the servlet accesses the SQL Server database using the JDBC/ODBC

Figure 5: How are access rights assigned to a user?

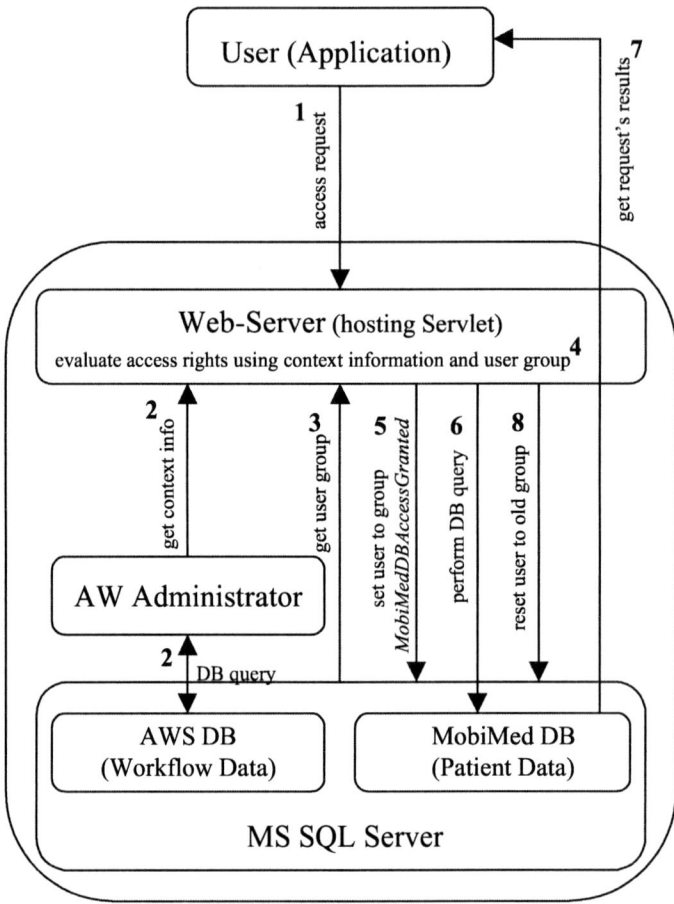

link.

From the SQL Server it receives user information as well as the access-control tables for the evaluation of access rights. The servlet handles the log-in of a user as well as setting and resetting her/him temporarily to the *MobiMedDBAccessGranted* group, enabling her/him to access data under the context-dependent access control scheme.

An example of a successful request under the context-dependent access control scheme is described in eight steps in Figure 5 (AW Administrator is the run-time environment of the workflow system):

1. The user (application), started by a user or triggered by the workflow system, sends an access request to the servlet.
2. The servlet requests and receives context information about the workflow of the patient whose record is to be accessed.
3. The servlet requests and receives group information about the user who

is requesting the access.

4. Both the context information and the role information are combined and the authority of the user to perform the access to the patient record is evaluated (using the access-control table). In the following it is assumed that the user has authority to perform the request (otherwise the request would be rejected at this point).
5. The user is set temporarily to group *MobiMedDBAccessGranted*.
6. The request (query) is performed on the database *MobiMedDB*.
7. The results of the query are delivered to the user.
8. The servlet resets the user to her/his previous group membership.

It is important to note that by no means is it possible for a user to access the database directly, since she/he does not belong to the group *MobiMedDBAccessGranted*.

In the first stage of the prototype implementation, user communication is handled by means of the servlet exchanging HTML documents with the browser of the user. In a later stage, this will be refined by implementing servlet/applet communication, allowing for more sophisticated user interfaces and more flexible applications.

DISCUSSION

We discuss in this section the security as well as the feasibility of the presented approach. To ensure privacy, the SQL Server and the Java Web server, including the access-control servlet, must be a single system. This means that if the database server and the Web server are not running on the same machine, they must, at least, be connected by an Intranet, which does not allow outside access. This is necessary, since Web server and database server communicate using JDBC/ODBC, which is not secure if the exchange of data can be seen from outside. The system is implemented in such a way that the Java Web server is the only component that can be accessed remotely, including the servlet it is hosting. Data exchange between client and server is held confidential by using SSL. The use of SSL has the additional benefit of supporting authentication of client and server. The client does not risk giving password information to a masqueraded server.

Given that the access-control system is implemented as described, the obvious major threat to the system is that the administrator log-in procedure known to the servlet becomes publicly accessible. By hosting the sensitive parts of the access control in a private method of the servlet and taking into account that servlets are server-sided Java bytecode, neither can a user access the private servlet methods nor is bytecode containing sensitive information accessible from the Internet, keeping this critical part of the system secure.

A way a user could try to break the system is by sending multiple access requests, knowing that at least for some of them she/he has authority. The idea is that a request for which she/he has no authority coincides sufficiently in time with a request for which she/he has authority. So both queries could be send to the SQL Server at the time when she/he is temporarily assigned to the group for which access is granted. This situation can only occur when multiple threads of the access-control servlet are not synchronized properly. The easiest attempt to tackle this problem is to disable multi-threading for single users.

Finally, is the presented approach really feasible? The only problem that could arise is that the servlet is not capable of analyzing an SQL query in order to obtain the information needed to evaluate access rights. We require that all access requests are specific to a given instantiation of the healthcare process, i.e., it is a specific instantiation for a particular patient. The access request can easily be analyzed to check whether it is a read-only access. Since information about the process state with respect to a patient is provided by the system and the system's user has authenticated herself/himself at log-in time, all necessary information to perform the context-dependent access control check is available to the system. Hence the proposed solution is indeed feasible. Since accesses to the system are relatively rare, compared to highly used Internet servers, performance aspects do not matter much. Tests with an early local prototype (Nitsche et al., 1998) did not show any problems with performance.

As a last comment, it should be understood that in the setting of a hospital, access to data can be vital. Therefore context-dependent access control will be equipped with simple, group-based bypass mechanisms. However, bypassing the context-dependent access control will have to be logged thoroughly to provide proof of potential misuse of the bypass mechanism.

CONCLUSION

In this chapter we reported on an ongoing project on context-dependent access control to support distributed clinical trials. We concentrated on presenting the technical aspects of the solution, in particular on the use of Java servlets. The implementation concept comprises a secure distributed solution to context-dependent access control over the Internet (Ultes-Nitsche & Teufel, 2000). The described system has not yet been implemented. It will be an extension to the current version of the MobiMed prototype that is presented in Röhrig and Knorr (2000). The new system will as well be PC-based (Windows NT/ Windows 2000), but using a Java servlet rather than a Perl script to access the MS SQL Server database (that contains the medical data) in a context-dependent fashion. To secure the communication the Secure Sockets Layer protocol (SSL) is used. The context information for the access control is supplied by a commercial workflow management system. Even though developed for a concrete platform, the underlying security concepts are general, and in particular the Internet security issues can be adapted to different platforms and applications (Hepworth & Ultes-Nitsche, 1999).

ACKNOWLEDGMENTS

We would like to thank Prof. Dr. Kurt Bauknecht and the Swiss National Science Foundation for their support as well as Konstantin Knorr for helpful suggestions on an early draft of this chapter.

ENDNOTES

[1] *MobiMed* (Privacy and Efficiency in *Mobile Med*ical Systems) is a project of the Swiss Priority Programme (SPP) for *Information and Communications Structures* (ICS, 1996-1999) of the Swiss National Science Foundation (SNSF) aiming at the development of mobile access to data in a clinical environment. Its contributing members are the Computer Science Department of the Universities of Zurich, Fribourg, and Southampton, and Plattner Schulz Partner AG, Basel.

[2] A health-care process is a business process describing patient treatment in a hospital. In our context it is the process a patient goes through in a clinical trial.

[3] API = Application Programming Interface.

REFERENCES

Fischer, H. R., Teufel, S., Muggli, C. and Bichsel, M. (1995). Privacy and Efficiency of Mobile Medical Systems (MobiMed). Bewilligtes Forschungsgesuch des SNSF Schwerpunktprogramms Informatikforschung SPP-IuK, Modul: Demonstrator, Nr. 5003-045359.

Freier, A. O., Karlton, P. and Kocher, P. C. (1996). The SSL Protocol Version 3.0. Internet Draft. November. Retrieved on the World Wide Web: http://home.netscape.com/eng/ssl3/draft302.txt. Netscape, Transport Layer Security Working Group.

Garfinkel, S. and Spafford, G. (1997). *Web Security and Commerce*. O'Reilly and Associates.

Hepworth, E. and Ultes-Nitsche, U. (1999). Security aspects of a Java-servlet-based Web-hosted e-mail system. *Proceedings of the 7th IFIP Working Conference on Information Security Management & Small Systems Security*. Boston: Kluwer Academic Publishers.

Holbein, R. (1996). *Secure Information Exchange in Organizations*. Doctoral thesis, University of Zurich, Switzerland. Aachen: Shaker Verlag.

Holbein, R. and Teufel, S. (1995). A context authentication service for role-based access control in distributed systems-CARDS. In Eloff, J. and von Solms, S. (Eds.), *Information Security-The Next Decade IFIP/SEC'95*. London: Chapman & Hall.

Holbein, R., Teufel, S. and Bauknecht, K. (1996). The use of business process models for security design in organisations. In Katsikas, S. and Gritzalis, D. (Eds.), *Information Systems Security-Facing the Information Society of the 21st Century (IFIP/SEC'96)*. London: Chapman & Hall.

Holbein, R., Teufel, S., Morger, O. and Bauknecht, K. (1997). A comprehensive need-to-know access control system and its application for medical information systems. In Yngström, L. and Carlsen, J. (Eds.), *Information Security in Research and Business (IFIP/SEC'97)*. London: Chapman & Hall.

Hunter, J. (1998). *Java Servlet Programming*. O'Reilly and Associates.

Medina-Mora, R., Winograd, T., Flores, R. and Flores, F. (1992). The action workflow approach to workflow management technology. *Proceedings of the Conference on Computer-Supported Cooperative Work (CSCW'92)*. ACM Press.

Nitsche, U., Holbein, R., Morger, O. and Teufel, S. (1998). Realization of a context-dependent access control mechanism on a commercial platform. In Papp, G. and Posch, R. (Eds.), *Global IT Security-Proceedings of the 14th International Information Security Conference (IFIP/Sec'98)*. Vienna: Austrian Computer Society, 116.

Röhrig, S., & Knorr, K. (2000). Towards a secure Web-based healthcare application. In Hansen, H. R., Bichler, M. and Mahrer, H. (Eds.), *A Cyberspace Odyssey-Proceedings of the 8th European Conference on Information Systems (ECIS'2000)*. Vienna: University of Economics and Business Administration.

Stallings, W. (1998). *Cryptography and Network Security*. New York: Prentice Hall.

Sun Microsystems. (1998). *Java Servlet API Whitepaper*.

Ting, T., Demurjian, S. and Hu, M. Y. (1991). Requirements, capabilities and functionalities of user-role-based security for an object-oriented design model. Landwehr, C. (Eds.). *IFIP WG 11.3 Workshop on Database Security*. Elsevier Science Publishers.

Ting, T., Demurjian, S. and Hu, M. Y. (1992). A specifcation methodology for user-role-based security in an object-oriented design model. In Thuraisingham, B. and Landwehr, C. (Eds.), *IFIP WG 11.3 6th Working Conference on Database Security*. Elsevier Science Publishers.

Ultes-Nitsche, U. and Teufel, S. (2000). Secure access to medical data over the Internet. In Hansen, H. R., Bichler, M. and Mahrer, H. (Eds.), *A Cyberspace Odyssey–Proceedings of the 8th European Conference on Information Systems (ECIS'2000)*. Vienna: University of Economics and Business Administration.

Chapter IX

POEMs in the Information Jungle—How Do Physicians Survive?

Allen F. Shaughnessy
Harrisburg Family Practice Residency Program, USA

David C. Slawson
University of Virginia Health Services Center, USA

Joachim E. Fischer
University Hospital of Zurich, Switzerland

INTRODUCTION

The days when newly graduated doctors were well equipped with the knowledge and information they would need during a lifetime are long since gone. Today's clinicians' knowledge becomes almost as rapidly outdated as the analysts' forecasts on the stock market. Tsunamis of new articles reporting scientific achievements flood the shorelines of current knowledge. Modern physicians need to be lifelong learners in order to adapt to the rapidly evolving medical environment. But how can physicians survive in the information jungle? What are the tools they need to weave a fabric of best medical practice that is woven from the relevant scientific knowledge and the detailed information about the patients' preferences? Are medical schools and the postgraduate educational systems preparing doctors for this?

LEARNING TO LEARN

When we began to acquire medical knowledge and competence as students, most of us either found teachers and professors who directed our learning by emphasizing the relevant information, or we prepared ourselves by learning according to required contents lists. Those of us who were good at "stocking their memory" made good medical students, excelled in school and performed well on tests. They became experts at recalling the right answer when the question came up in the examination.

During medical apprenticeship as residents we gained proficiency in the basic mechanical skills. However, most of us experienced that the memorized lists of differential diagnoses and related symptoms were of little help in identifying our patients' relevant problem. We relied on the seniors' experience to find a path to diagnosis and treatment. Our university-based knowledge equipped us well to find a way from the trunk of a diagnostic tree to its outmost branches, but we were often unable to identify the roots of a problem when the only traces were a few symptoms. Like strangers in the jungle, we were unable to read from fallen leafs.

INFORMATION OVERLOAD

As doctors advance in their medical career, the less valuable becomes the knowledge acquired in medical school. Many become frustrated on this rapid decay of excellence and expertise. Because they are at a different point of executing medical skills, the invited speakers at the congresses worldwide do no longer substitute for the professors or teachers in medical school. Only few have learned effectively how to learn and how to weed through the overwhelming amount of new information (Argyris, 1991; Hamm & Zubialde, 1995; Weed, 1997). With the average reading time of less than one hour per week, it is simply impossible to keep up by reading the journals of one's specialty. Moreover, some of the most relevant articles in the field often get published outside the specialty in one of the more prestigious general journals. The task to garden one's knowledge is like keeping a fruit garden in the midst of a jungle. No one is there to direct doctors towards the new information they need and no one helps to identify outdated knowledge that should be weeded (Goodwin & Goodwin, 1984; Hills, 1993).

Some defend themselves against the information overload by setting up fierce barriers around their existing and accumulated knowledge. These barriers defend the "personal experience" like walls around medieval castles. These walls are a major source of prejudices and biases and may effectively shut out new ideas (Hills, 1993).

Problem-based learning has been suggested as a remedy to this. Graduates from medical schools focussing on problem-based learning appear to be better equipped with the tools for lifelong learning than those of a traditional curriculum (Norman & Schmidt, 1992; Shin, Haynes & Johnston, 1993). Probably, early medical education induces a lifelong attachment to specific patterns of learning. If this is the traditional "lecture and test" method of teaching, young doctors leave the university without the skills enabling them to constantly update and replace this knowledge. One result is the failure of continuing medical education to change the actions of doctors (Cantillon & Jones, 1999).

PARADOX OF ADULT LEARNERS

An under-recognized problem is the paradox of adult learning. One of the distinctions between adults and children is the freedom of choice. Adults decide what to learn. However, when adults who were conditioned by the traditional type of "store and retrieve" learning have to acquire new knowledge, they rapidly revert back to the child-mode of passive dependency on the teacher. However, the deeper need to be self-directed may surface again, reasserting itself and creating an unsolvable situation for the educator (Knowles, 1984).

The economic consequences of the medical education system's failure to provide adequate tools for lifelong learning are incalculable. It has been estimated that in the USA "180,000 people die each year partly as a result of iatrogenic injury, the equivalent of three jumbo jet crashes every two days" (Leape, 1994). The main reason for this is human error: errors of judgment, relying on the human brain's limited capacity to consider multiple variables, and error from applying outdated knowledge.

The inability of the traditional pedagogical techniques to improve on physicians' skills and to further lifelong learning has long been recognized (Ellrodt et al., 1997; General Medical Council, 1993; Neufeld & Barrows, 1974). Much hope has been vested with the new information technologies. Today, computers facilitate rapid access to database storehouses and potentially deliver information at everyone's fingertips. But although the majority of physicians own a computer and many regularly surf the Internet, few actually use the electronic device in patient management decisions. (Koller, Grütter, Peltenburg, Fischer & Steurer, 2001). The majority of Internet users consider information retrieval time-consuming and cumbersome. Probably, too little attention has been given to the actual delivery process.

DELIVERING THE INFORMATION PRODUCT

Much could be learned from the production of merchandisable goods. A major limitation to this analogy should, however, be recognized upfront. Currently, few physicians are willing to pay for information (Koller et al., 2001). But production of information (e.g., conducting a multicenter trial), aggregation of information to knowledge (e.g., by means of a Cochrane review), storing this information in a database and the preparation for delivery (e.g., setting up a Web site) are all highly cost-intensive production steps. Even on the global information market, there is no free lunch.

In order to deliver the best information into the hands of the doctor (and the patient) at the point of care on time, four distinct components are required (see Figure 1):

• Production of information: clinical experience and original research to produce the crude data.

• Refinement: The raw data must be processed and refined for use. In the medical information process, the first step is the aggregation of the data to a publication in a peer-reviewed journal. Further refinement occurs through synthesizing information from several original reports. The products of this secondary process may be systematic reviews and meta-analyses, cost-effectiveness analyses, or guidelines. Undoubtedly the past decades have witnessed an enormous

Figure 1: The information process

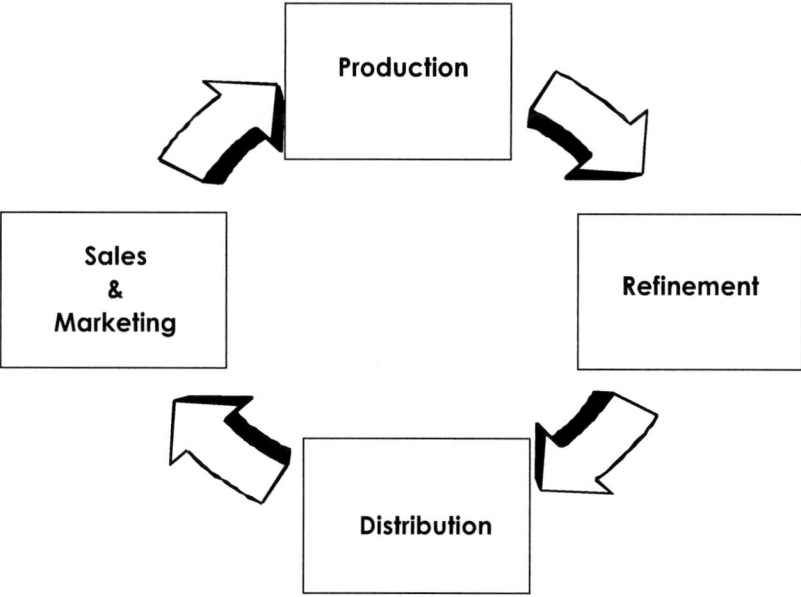

progress both in quantity and quality as to these first two steps in the production of medical information.

- Distribution: In stark contrast, little progress has been made in the delivery of information. More than 500 years ago, Gutenberg invented a method to spread information that was hitherto confined to scholars. It appears that in the hands of modern physicians, well-presented textbooks are still esteemed as much as the printed Bible half a millennium ago (Koller et al., 2001). Hopefully, this preference for the written word instead of the displayed word is just a question of maturation of the new technologies. End users of books are probably not aware of the many little details like choice of fonts, colours, paper quality, or binding that make the difference between a fringe brochure and a good book. If the modern technology would be delivering information by point and click to bright, 2000 x 2000 pixel, large-format TFT screens at the speed of flicking a page in the textbook, and the design and organization of books would be still confined to the state of the 16th century, undoubtedly end users would embrace the new information technology wholeheartedly.

- Finally, the sales and marketing of medical information has largely been ignored. In the merchandise industry, manufacturers tailor products and marketing to the needs of the users (this may be a nonconscious need, e.g., products of the tobacco industry). In contrast, the end users of medical research are left to stumble on new information on their own. The consequence is that many new research findings in medicine fail to reach the consumers who need them the most (doctors and patients). From the real end user's (the patients') perspective, valid new information remains being stored in remote databases like scientific knowledge in medieval libraries. This constitutes the paradox of the information age: Lack of information at the point of care within a context of information overload. The situation is like in a jungle with its thousands of plants, of which most are of little use to humans and many are poisonous. The key to survival is the knowledge to identify the few fruits that nourish humans.

EVIDENCE-BASED MEDICINE

Evidence-based medicine has been welcomed as the long awaited remedy to these problems. The protagonists of evidence medicine claim to follow an approach to practice and teaching that integrates pathophysiological rationale, caregiver experience, and patient preferences with valid and current clinical research evidence (Ellrodt et al., 1997). Simply put, this new school of thought promises to provide a bridge between practice and best available evidence from contemporary research when making decisions about the care of individual patients. But the problems related to lifelong learning start with the first steps in practicing evidence-based medicine: The need to precisely define a clinical question, the successful search for information and the critical appraisal of the content. Theoretically, physicians should be armed with the best evidence for clinical decisions (Ellrodt et al., 1997). But in daily practice, these tools and techniques are too clumsy and time consuming. Few clinicians are able to distill an answerable question from a current clinical problem, type the request into a computer and come up with a reasonable answer in less than 15 minutes. Therefore, it is not surprising that few clinicians use the evidence-based medicine approach in its pure form (McAlister, Graham, Karr & Laupacis, 1999).

GETTING RESEARCH INTO PRACTICE

Even if busy doctors regularly hunt for information as it arises from problems in the practice, this problem-driven mode of information gathering of traditional evidence-based medicine does not provide for alerts to important new evidence. Unless doctors are actively alerted to new information outdating their current knowledge, it is likely that important changes escape their attention. To achieve the aim of providing best information at the point of care requires additional input. Simple monitoring of the specialty journals for being alerted can become a dangerous practice if the information is not reviewed within its context and considering other conflicting data (Slawson, Shaughnessy & Bennett, 1994). Therefore, it is mandatory to keep busy clinicians updated with the continuing evolution of the available evidence on the subject. Attempts to this extent are realized by the Cochrane collaboration, which encourages regular update of reviews.

But even the best and regularly updated evidence often does not change clinical practice. It has been shown that physicians often rely on other sources of information than research or its synthesis in the pure form (McAlister et al., 1999). Observational studies on information gathering revealed that other sources, from pharmaceutical representatives to trustworthy colleagues, are more commonly used (Ely, Levy & Hartz, 1999; Gruppen, Wolf, Voorhees & Stross, 1987; Smith, 1996).

PATIENT-ORIENTED EVIDENCE THAT MATTERS

A simple rule of marketing probably governs the adult learning of doctors: New information will only be retained if there is a "need to know" (Knowles, 1984). Three criteria characterize potentially important and relevant evidence, which should be delivered at the point of care. First, the evidence focuses on outcomes relevant to the patient. While patients have little primary interest in the composition of their blood fat, they all share the desire to live long, pain-free and functionally rewarding lives. Second, the new information addresses a problem that is common to the doctor's practice. Third, the information has the potential to cause physicians to change their behaviour.

To label such important and relevant information, the acronym "POEMs" (patient-oriented evidence that matters) has been created (Slawson et al., 1994). A potential strategy to survive in the information jungle is to always have available two tools: a *foraging* and a *hunting* tool. Every health discipline should have its own specialty-specific "POEM bulletin board" to alert clinicians of new valid POEMs on common issues as they become available. Second, all clinicians must employ a hunting tool when in the mode of problem-driven information gathering (when searching for answers to specific patient problems) that includes the cumulative POEM database specific to each specialty.

Innumerable Web sites and medical portals aim to provide for this continuous alerting and update by newsletters, bulletin boards and other systems. The inherent problem of all these tools is the potential bias that is introduced by the person who filters the information prior to distribution. Few of the services provide information on the criteria for relevance and validity, if such filters are employed at all (McKibbon, 1998; Nutting, 1999).

A further unsolved problem is the management of this information. Clinicians not only need tools for retrieving the information rapidly, but also ways to manage the updated information within their minds (Jacobson, Edwards, Granier & Butler, 1997). Well-designed computer based programs allow one to retrieve information within less than 30 seconds (Straus, 1999). Thus, rather than aiming that physicians keep storing the information in their minds, modern knowledge management techniques should emphasize memorization of flags and paths to information retrieval. Probably, next generations' physicians will use mobile networking and handheld microcomputers, replacing the small paper reference notebooks many clinicians carry in their pockets today to retrieve information at the point of care (Ebell & Barry, 1998).

HUNTER AND FORAGER IN THE INFORMATION JUNGLE

Medical schools provide future doctors with some means to structure the basal knowledge required to practice the profession. The skills learned from the traditional curriculum probably work well during residency. However, when released into the real world of being responsible for patient care, other tools are needed to survive in the information jungle that are currently not taught in medical school. Physicians must be expert hunters and foragers for relevant medical information. To the benefit of the patient it is essential that doctors are capable of finding, evaluating, and applying new information as it becomes available.

SUMMARY POINTS

- Traditional university-based teaching methods equip students with vast amounts of information but they fail to teach the skills for continuous learning.
- While computers and the Internet have made information readily available at everyone's fingertips, little consideration has been given to how information is delivered.
- The primary focus of evidence-based medicine is to identify and validate written information; for most doctors this is a too time-consuming process.
- If best available evidence is to be used at the point of care, sophisticated filters are needed that increase the yield on relevance of the information.
- Doctors need an alert method for becoming aware of relevant new information that implies the need to update their knowledge; these systems should be tailored to the doctors' individual needs.

This chapter is based on a paper titled "Are we providing doctors with the training and tools for lifelong learning?" by A. F. Shaughnessy and D. C. Slawson published in May-June 1999 in the *British Medical Journal*, Vol. 319, pp. 1280-1283.

REFERENCES

Argyris, C. (1991). Teaching smart people how to learn. *Harvard Business Review*, May-June, 99-109.

Cantillon, P. and Jones, R. (1999). Does continuing medical education in general practice make a difference? *British Medical Journal*, 318, 1276-1279.

Ebell, M. H. and Barry, H. C. (1998). InfoRetriever, rapid access to evidence-based information on a hand-held computer. *MD Comput*, 15, 289, 292-7.

Ellrodt, G., Cook, D. J., Lee, J., Cho, M., Hunt, D. and Weingarten, S. (1997). Evidence-based disease management. *JAMA*, 278, 1687-1692.

Ely, J. W., Levy, B. T. and Hartz, A. (1999). What clinical information resources are available in family physicians' offices? *J Fam Pract*, (48), 135-139.

General Medical Council. (1993). *Tomorrow's Doctors*. Recommendations on undergraduate medical education. London, GMC.

Goodwin, J. S. and Goodwin, J. M. (1984). The tomato effect: Rejection of highly efficacious therapies. *JAMA*, 251, 2387-2390.

Gruppen, L. D., Wolf, F. M., Voorhees, C. V. and Stross, J. K. (1987). Information-seeking strategies and differences among primary care physicians. *Mobius*, 71, 18-26.

Hamm, R. M. and Zubialde, J. (1995). Physicians' expert cognition and the problem of cognitive biases. *Primary Care*, 22, 181-212.

Hills, G. (1993). The knowledge disease. *British Medical Journal*, 307, 1578.

Jacobson, L. D., Edwards, A. G., Granier, S. K. and Butler, C. C. (1997). Evidence-based medicine and general practice. *British Journal of General Practice*, (47), 449-452.

Knowles, M. S. (1984). Introduction, the art and science of helping adults learn. In *Androgogy in Action. Applying Modern Principles of Adult Learning*. San Francisco: Jossey-Bass, 1-21.

Koller, M., Grütter, R., Peltenburg, M., Fischer, J. E. and Steurer, J. (2001) Use of the Internet by Medical Doctors in Switzerland. *Swiss Med Weekly*, 131, 251-254.

Leape, L. (1994). Error in medicine. *JAMA*, 272, 1851.

McAlister, F. A., Graham, I., Karr, G. W. and Laupacis, A. (1999). Evidence-based medicine and the practising clinician. *Journal of General Intern Medical*, 14, 236-242.

McKibbon, K.A. (1998). Using 'best evidence' in clinical practice. *ACP Journal Club*, 128, A15.

Neufeld, V. R. and Barrows, H. S. (1974). The "McMaster philosophy", an approach to medical education. *Journal of Medical Education*, 49, 1040-1050.

Norman, G. R. and Schmidt, H. G. (1992). The psychological basis of problem-based learning, a review of the evidence. *Academy Medical*, 67, 557-565.

Nutting, P. A. (1999). Tools for survival in the information jungle. *Journal of Family Practice*, 48, 339-340.

Shin, J. H., Haynes, R. B. and Johnston, M. E. (1993). Effect of problem-based, self-directed undergraduate education on life-long learning. *Can Medical Association Journal*, 148, 969-976.

Slawson, D. C., Shaughnessy, A. F. and Bennett, J. H. (1994). Becoming a medical information master, feeling good about not knowing everything. *Journal of Family Practice*, 38, 505-513.

Smith, R. (1996). What clinical information do doctors need? *British Medical Journal*, 313, 1062-1068.

Straus, S. E. (1999). Bringing evidence to the point of care. *Evidence-Based Medicine*, 4, 70-71.

Weed, L. L. (1997). New connections between medical knowledge and patient care. *British Medical Journal*, 315, 231-235.

Chapter X

The Medical Journal Club— A Tool For Knowledge Refinement and Transfer In Healthcare

Khalid S. Khan
Birmingham Women's Hospital, United Kingdom

Lucas M. Bachmann
Birmingham Women's Hospital, United Kingdom
University Hospital of Zurich, Switzerland

Johann Steurer
University Hospital of Zurich, Switzerland

INTRODUCTION

The information base for healthcare is rapidly expanding. There are more than 20,000 biomedical journals. Approximately 17,000 new biomedical books were published in 1990, and these were projected to increase by annually up to 7% (Sackett et al., 1996; Siegel, Cummings, & Woodsmall, 1990). With such an exponential increase in information, there is a need for effective and efficient strategies to keep up-to-date with clinically relevant new knowledge. Without current best evidence, medical practice risks becoming out-of-date, to the detriment of patients.

In the last decades it has been accepted, that knowledge management strategies are required for physicians after they have finished their formal undergraduate and postgraduate training (Davis et al., 1999) to maintain and to improve clinical performance (Cantillon & Jones, 1999). These have often amalgamated under the heading of continuing medical education (CME). Davis defines CME as "any and all ways by which doctors learn after the formal completion of their training" (Davis & Fox, 1994). In the United States, physicians' time spent in continuing education activities has been reported to exceed 50 hours a year (Davis et al., 1999).

In 1999 Davis and coworkers analyzed the evidence on the effectiveness of traditional CME activities on physicians' performance and healthcare outcomes. They found that only interaction and enhanced participant activity improves professional practice and may lead to improvements in healthcare. Cantillon and Jones (1999) suggest that CME should be based on physicians' daily work and that CME activities should include peer review group based learning. Nevertheless, the traditional style of authority-led clinical practice and teaching still dominates CME activities. We take the example of the medical journal club, a scientific meeting for dissemination of new medical knowledge, which has usually been based on the traditional approach described above. In this chapter we analyze the effects of this traditional approach on knowledge management and propose ways to improve knowledge refinement and transfer by a new approach to the medical journal club.

KNOWLEDGE MANAGEMENT IN TRADITIONAL MEDICAL JOURNAL CLUBS

An analysis of traditional journal club meetings shows that the participants or members are left to select and appraise the articles with little or no structure and often no clear purpose (Khan & Gee, 1999). Very often the participant picks up articles arbitrarily or gets them from an authority without any systematic literature search and evaluates them without explicit methods of appraisal (Rosenberg & Donald, 1995). These include evaluation of the validity and applicability of the findings of the article (Oxman, Sackett, & Guyatt, 1993). Most doctors do not appraise the quality of the studies because it is generally assumed that the journal editors and reviewers would have appraised the manuscript prior to publication. This assumption is fraught with some difficulty, as 80–90% of published literature in leading journals does not fulfill basic criteria for validity (Haynes, 1993). Often the presentation consists of a summary of research procedures, results and conclusions without appraisal of validity and applicablility.

Without critical appraisal of the information being presented, the journal club becomes a session in presentation skills. The presenters often end up receiving criticism by other participants or mentors for not appropriately selecting or appraising the articles. Mentors' comments are generally based on personal experience and opinion, which is usually methodologically unsound. This may be because the mentors themselves do not have skills in critical appraisal of the medical literature. Other members of the journal club who attend remain passive and they may not gain any additional knowledge. As a result, they hardly feel enthused to take the role of a presenter in subsequent journal club sessions. The findings of the presentation are neither gathered systematically nor stored in a manner that they can be offered to other members in the same healthcare area. An important opportunity for knowledge refinement and knowledge transfer is missed.

CHANGING TRADITIONAL KNOWLEDGE MANAGEMENT STRATEGIES

To address the above problems in knowledge management, modern continuing education and postgraduate training programs have started to define the role of the journal club in terms of objectives and learning outcomes. The most common objectives of the journal club have been teaching perspective throughout critical appraisal of the medical literature, to help keep up-to-date with the literature, to teach research design and to have an impact on clinical practice (Alguire, 1998). Many journal clubs have attempted to incorporate the recent trends in teaching to facilitate learning, for example, by adoption of approaches based on active learner participation (Inui, 1981) and problem-based learning (Joorabchi, 1984). However, the efforts to improve the educational value of journal clubs in terms of trainees' learning outcomes have been limited (Alguire, 1998; Sidorov, 1995), and controlled trials assessing the value of journal clubs for teaching critical appraisal skills have shown only moderate improvements (Norman & Shannon, 1998).

In the early 1980s David Kolb presented a visualization of experiential learning (1984). The theory describes a learning cycle in which knowledge may be formed and adapted through the learning experience. The concept consists of learners' involvement in new experience, reflection on the experience, conceptualization of the concrete experience into a logical framework and testing the new framework in decision making and problem solving (Figure 1).

At the same time, new concepts about how medical knowledge should be interpreted and acquired have consolidated under the umbrella of evidence-based medicine (Evidence-Based Medicine Working Group, 1992). The term evidence-based medicine, or EBM as it has come to be known popularly, was coined at McMaster Medical School in Canada in the 1980s to name the clinical strategy of asking questions, finding and appraising relevant data, and harnessing this information for everyday clinical practice, an approach that the school had been developing for over a decade (Rosenberg & Donald, 1995). Technically, EBM can be defined as:

"An approach to health care that promotes the collection, interpretation, and integration of valid, important and applicable patient reported, clinician observed, and research derived evidence. The best available evidence, moderated by patient circumstances and preferences, is applied to improve the quality of clinical judgments."
(McKibbon, Wilczynski, Hayward, Walker, & Haynes, 1996)

In the knowledge context however, EBM represents two separate but related philosophical entities: a new school of medical epistemology and a new paradigm of knowledge-based clinical practice (Tonelli, 1998). The EBM epistemology defines the nature of knowledge, giving its users the basis for knowing and the validity of different ways of knowing. It provides hierarchies of evidence to determine what constitutes best available evidence. EBM considers clinically relevant patient-centered research as evidentiary. It generally considers physiologic principles, expert opinion

Figure 1: Kolbs Learning Cycle. Adapted from Fry, Ketteridge and Marshall (1999)

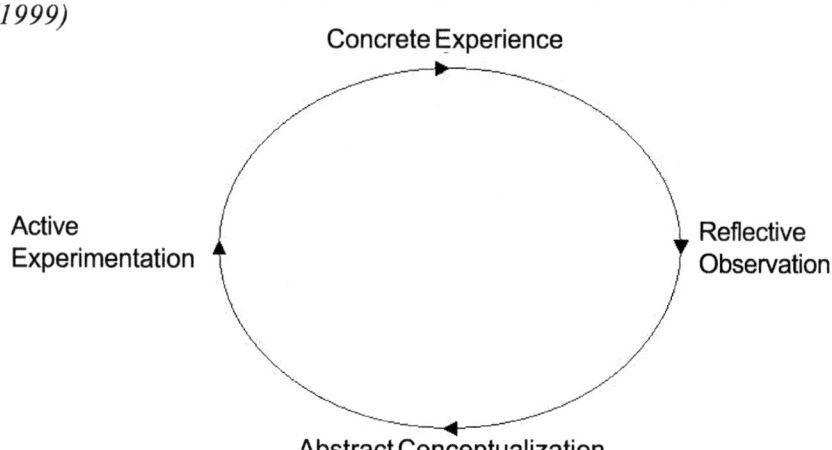

and case-based experiences as non-evidentiary. This epistemological framework forms the basis on which the knowledge-based clinical practice paradigm is built. The EBM paradigm focuses on promoting the use of evidentiary knowledge in medical decision-making and attempts to minimize the use of non-evidentiary knowledge. This represents a break away from traditional practice whose roots often lay in knowledge regarded as non-evidentiary by EBM.

Interestingly, it was not until 1992 that this clinical strategy officially appeared in a peer-reviewed journal, the *Journal of the American Medical Association* (Evidence-Based Medicine Working Group, 1992). Central to the EBM concept is critical literature appraisal skills, which refers to the process of evaluating the evidence for its validity or quality (Rosenberg & Donald, 1995), thereby classifying knowledge as evidentiary or non-evidentiary. High quality evidence is expected to provide new, more powerful, more accurate, more efficacious, and safer means of diagnosis and therapy for patients (Sackett et al., 1996). However, EBM is often misunderstood by practicing clinicians and this is partly because EBM has traditionally not been covered in undergraduate education. The origins, rationale and limitations of EBM are often unclear to many clinicians and other healthcare staff. Currently, EBM principles and skills are often taught during postgraduate and continuing education. However, there is much to be desired from such activities in terms of knowledge management (Khan, 2000).

In order to enhance the value of journal club in knowledge management, a new format, which departs from a traditional approach, is required (Khan & Gee, 1999). To bring about familiarity with the principles of EBM and to acquire new knowledge, appraise and apply it in clinical practice, the new format incorporates recent trends in medical education where the aim of teaching is to facilitate learning. In particular educational programs should inculcate deep learning. This is because, in contrast to superficial learning, which circles around memorizing and reproducing and forgetting soon afterwards, the deep approach helps doctors make sense out of the subject matter (Brown & Atkins, 1988; Gibbs, 1992). In accordance with Kolb's suggestions (1984), the deep approach is fostered when the process of learning builds on activation of existing knowledge, construction of new knowledge over and above what is existing, and refinement of the newly acquired knowledge. To achieve this objective in the journal club a clinical context is required which motivates participants to take an active role and to interact with others in the learning process (Cantillon & Jones, 1999; Davis et al., 1999; Gibbs, 1992).

KNOWLEDGE MANAGEMENT IN NEW MEDICAL JOURNAL CLUBS

It is possible to organize journal clubs based on a guided discovery approach, which puts the learning process in the context of clinical problem solving (Khan et al., 1999). There are several steps to the proposed new format (Figure 2).

As a first step, the participants of the journal club should be encouraged to identify a current patient care problem that they consider important but about whose management they feel uncertain, possibly in the light of conflicting expert opinion. They bring out issues which they feel are worth the effort of exploring the medical literature to address their uncertainty. The problem formulation process is to be led by doctors themselves (Sidorov, 1995), in light of the evidence that journal clubs independent of faculty are more successful.

The second step is to search the literature to identify relevant articles. Facilities for computerized literature searches of electronic bibliographic databases are provided in most postgraduate education resource centers and on the Internet. This process should be structured and conducted using key words representing the population, interventions, and outcomes of interest. Often an EBM search filter for therapy, diagnosis, prognosis or etiology (Haynes et al., 1994) limits the search. The journal club members collectively decide on article/articles that are potentially relevant after going

Figure 2: Four steps of the proposed new format: identification of a current patient care problem, asking a research question, finding and appraising relevant data, and harnessing the information for everyday clinical practice.

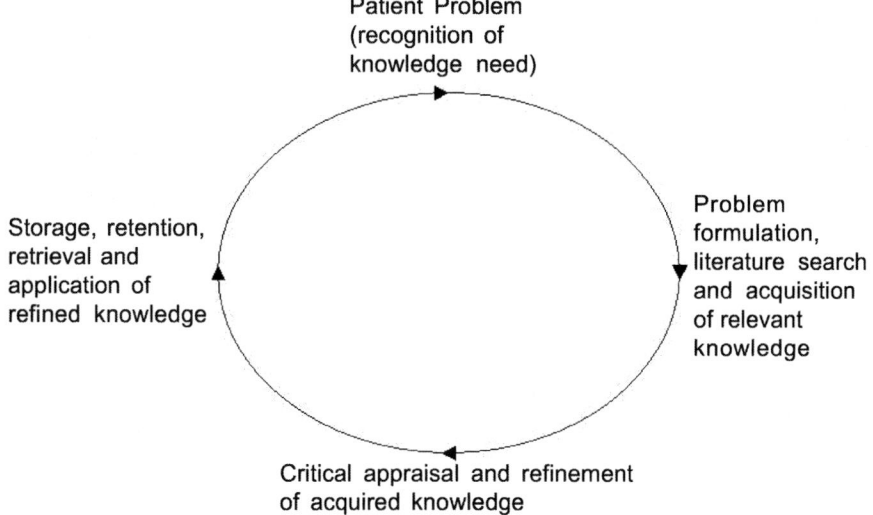

through the citation list. These articles are retrieved from the base library or ordered through the interlibrary loan system.

Having acquired the identified articles, participants appraise them independently at their own pace. The appraisal is based on structured guidelines, as this approach is known to enhance the value of journal clubs (Burstein, Hollander, & Barlas, 1996). The structured appraisal is facilitated by use of computer software that employs few simple criteria for methodological rigour. This software is used to perform the calculations of clinically meaningful measures of effect and allows the appraisal to be stored in an electronically retrievable form (Badenoch, Straus, & Sackett, 2000). The doctor and the mentor together refine the electronic appraisal, decide on the format of the journal club presentation and produce instructional materials. At the end of the presentation the presenter gets feedback and has a chance for reflection. The appraisal is finally modified in light of the discussion in the journal club meeting and a final version of the appraisal can be stored electronically in any database. The incorporation of computer technology in the journal club leads to a completely new dimension of knowledge refinement and transfer.

The appraised topics can be retrieved from this electronic bank for use in clinical service areas, future teaching and ward rounds. Once a database is built, instant retrieval of information abstracted from the appraisal of the article in this manner can be applied to resolve patient management problems, which reinforces the impact of problem-solving skills learned during preparation for the journal club. In addition, the aspect of documentation provides an incentive for the presenters to disseminate their work to others who were unable to attend the meeting. The hereby presented new format has been proposed recently (Khan & Gee, 1999) and has been shown to be effective in a clinical setting (Khan et al., 1999).

IMPLEMENTING THE NEW JOURNAL CLUB

The theory of the new journal club and its educational approach are attractive but the implementation can pose several problems. It is important to appreciate the value of clinical expertise of the senior clinicians. By actively involving senior clinicians in the process, one can convince them that the journal club is a worthwhile exercise. Moreover, by exposing the medical staff to the added value of new evidence in the light of existing experience, one can demonstrate the relevance of the approach. One should recognize that there are no quick and easy knowledge-based answers to many clinical problems. Appraisal of the selected literature is a vital component of the journal club.

Lack of skills in critical appraisal can be overcome by holding workshops. Active participation of the clinical librarian will facilitate searching and retrieval of articles. Many doctors have limited computing skills. Having a core group of participants who help others to use the computers effectively for knowledge refinement can obviate this problem.

Enthusiasm of participants, introduction of computer technology and integration of senior clinicians' experience are bound to make the new format journal club a success in knowledge management in a clinical setting (Khan et al., 1999).

KEY POINTS

1. With the enormous, rapid and exponential expansion in the medical literature, there is a need for effective and efficient strategies to keep abreast of relevant new knowledge.
2. The medical journal club, traditionally used as an educational tool in postgraduate and continuing medical education, can be designed for acquisition and appraisal of relevant, current best clinical evidence by using a systematic approach to both acquisition and appraisal of evidence in a context directly related to patient care.
3. Incorporation of computer technology in the journal club helps with acquisition, appraisal and refinement of knowledge, but more importantly it allows for knowledge transfer by making possible the storage and instant retrieval of appraised topics in the future.

A NEW KNOWLEDGE MANAGEMENT STRATEGY FOR THE MEDICAL JOURNAL CLUB

Aim

To facilitate the use of new relevant knowledge from the medical literature in guiding clinical decision-making.

Objective

To prepare journal club members to identify, appraise and present in turn published articles concerning etiology, diagnosis, prognosis and therapy of patient problems seen in day-to-day clinical practice.

Knowledge Identification, Storage and Retrieval Strategy

For each journal club session the presenter is given a clinical question about diagnosis, therapy, etc. should:

- Frame clinical questions in an answerable form.
- Conduct a search for electronic bibliographic databases to identify relevant articles.
- Critically appraise the article for validity, importance of results and clinical applicability.
- Record the above information in an electronic form using computer software for storage in an electronic bank.
- Retrieve the appraised knowledge from the electronic storage site for clinical decision-making.

Knowledge Refinement Strategy

For each journal club presentation, the teaching and learning strategy described below allows knowledge refinement to take place:

- Deciding how to respond to a clinical scenario (small group discussion)
- Understanding the methodological guidelines for critical appraisal (independent learning, one-to-one tutoring and peer tutoring)
- Appraisal of research paper using the methodological guidelines (independent learning, one-to-one-tutoring and peer tutoring)
- Deciding how to respond to the findings of research article (small group discussion).

REFERENCES

Alguire, P.C. (1998). A review of journal clubs in postgraduate medical education. *Journal of General Intern. Medical*, 13, 347-353.

Badenoch, D., Straus, S. and Sackett, D. (2000). Catmaker 2000. Oxford, *NHS Research and Development Centre for Evidence Based Medicine*.

Brown, G. and Atkins, M. (1988). Studies of student learning. In A.M. Brown, G. (Ed.), *Effective Teaching in Higher Education*, 150-158. London: Routledge.

Burstein, J. L., Hollander, J. E. and Barlas, D. (1996). Enhancing the value of journal club: Use of a structured review instrument. *American Journal of Emergency Medication*, 14, 561-563.

Cantillon, P. and Jones, R. (1999). Does continuing medical education in general practice make a difference? *British Medical Journal*, 318, 1276-1279.

Davis, D., O'Brien, M. A., Freemantle, N., Wolf, F. M., Mazmanian, P. and Taylor-Vaisey, A. (1999). Impact of formal continuing medical education: do conferences, workshops, rounds, and other traditional continuing education activities change physician behavior or health care outcomes? *JAMA*, 282, 867-874.

Davis, D. A. and Fox, R. D. (1994). *The Physician as Learner: Linking Research to Practice.* Chicago, IL: American Medical Association.

Evidence-Based Medicine Working Group. (1992). Evidence-based medicine. A new approach to teaching the practice of medicine. *Journal of American Medical Association*, 268, 2420-2425.

Fry H., Ketteridge S. and Marshall S. (1999). Understanding Student Learning. In Fry, H., Ketteridge S. and Marshall S. (Eds.), *A Handbook for Teaching & Learning in Higher Education*, 21-40. London: Kongan Page.

Gibbs, G. (1992). The nature of quality of learning. In Gibbs, G (Ed.), *Improving the Quality of Student Learning*, 1-11. Technical and Educational Services Ltd.

Haynes, R. B. (1993). Where is the meat in clincial journals? *ACP Journal Club.*

Haynes, R. B., Wilczynski, N., McKibbon, K. A., Walker, C. J. and Sinclair, J. C. (1994). Developing optimal search strategies for detecting clinically sound studies in MEDLINE. *Journal of American Medical Information Association*, 1, 447-458.

Inui, T. S. (1981). Critical reading seminars for medical residents. Report of a teaching technique. *Medical Care*, 19, 122-124.

Joorabchi, B. (1984). A problem-based journal club. *Journal of Medical Educucation*, 59, 755-757.

Khan, K., Pakkal, M., Brace, V., Dwarakanath, L. and Awonuga, A. (1999). Postgraduate journal club as a means of promoting evidence-based obstetrics and gynaecology. *Journal of Obstetrics and Gynaecology*, 19, 231-234.

Khan, K. S. (2000). *Teaching Evidence-based Medicine in Postgraduate and Continuing Education Curriculum Design and Assessment* (Unpublished master's thesis).

Khan, K. S., and Gee, H. (1999). A new approach to teaching and learning in journal club. *Medical Teacher*, 21, 289-293.

Kolb, D. (1984). *Experiential Learning*. New Jersey: Prentice-Hall.

McKibbon, K., Wilczynski, N., Hayward, R., Walker, C. J. and Haynes, R. B. (1996). *The Medical Literature as a Resource for Evidence-Based Care* (Working Paper. McMaster University, Health Information Research Unit, McMaster University, Ontario, Canada.

Norman, G. R. and Shannon, S. I. (1998). Effectiveness of instruction in critical appraisal (evidence-based medicine) skills: A critical appraisal [see comments]. *CMAJ*, 158, 177-181.

Oxman, A. D., Sackett, D. L. and Guyatt, G. H. (1993). Users' guides to the medical literature. I. How to get started. The Evidence-Based Medicine Working Group. *JAMA*, 270, 2093-2095.

Rosenberg, W. and Donald, A. (1995). Evidence based medicine: an approach to clinical problem-solving [see comments]. *British Medical Journal*, 310, 1122-1126.

Sackett, D. L., Rosenberg, W. M., Gray, J. A., Haynes, R. B. and Richardson, W. S. (1996). Evidence based medicine: what it is and what it isn't [editorial] [see comments]. *British Medical Journal*, 312, 71-72.

Sidorov, J. (1995). How are internal medicine residency journal clubs organized, and what makes them successful? *Arch.Intern.Med.*, 155, 1193-1197.

Siegel, E., Cummings, M. and Woodsmall, R. (1990). Bibliographic-retrieval systems. In Shortliffe, E. H. (Ed.), *Medical Informatics: Computer Applications in Health Care*. Reading, MA: Addison-Wesley Publishing & Co.

Tonelli, M. (1998). The philosophical limits of Evidence-based medicine. *Academic Medicine*, 73, 1234-1240.

Chapter XI

Using On-Line Medical Knowledge To Support Evidence-Based Practice: A Case Study of a Successful Knowledge Management Project[1]

Daniel L. Moody and Graeme G. Shanks
University of Melbourne, Australia

INTRODUCTION

The Problem of Medical Decision-Making

Making appropriate decisions about patient care relies on making connections between:

- *Information about the patient's condition*: This includes pathology tests, X-rays, observations, examinations and medical records.
- *Medical knowledge*: this includes knowledge about diseases, their symptoms, causes and treatments.

To do this effectively, clinicians need to take into account all relevant medical research and integrate it with detailed data about the patient's condition. This is an extremely difficult task, not least because it is estimated that the amount of medical knowledge doubles every five years (Weed, 1997). Staying abreast of the latest developments in medical research therefore represents an enormous intellectual challenge for clinicians. Faced with information overload, doctors often fall back on global judgements based on experience rather than thorough analysis of the relevant medical literature. As a result, medical practice is surprisingly anecdotal and experience-based rather than being based on scientific fact and empirical results.

Knowledge Management

Over the last decade, there has been a major shift in the world economy from the production of goods to the provision of services. This has been hailed as a transition from an industrial economy to a *knowledge economy* (Sveiby, 1997). Managing knowledge has become a major concern in many organizations today and is increasingly being seen as a source of sustainable competitive advantage (Broadbent, 1997; Hansen et al., 1999; Nonaka & Takeuchi, 1995). While there are many different approaches to knowledge management, their objectives are the same: to make more effective use of "know-how" and expertise in an organization (Martin, 1999).

Knowledge is a high value form of information that can be used to make decisions and take action (Davenport et al., 1998). A key difference between knowledge and information or data is that it is intellectually intensive rather than IT-intensive; knowledge is the result of human interpretation and analysis rather than data processing. Knowledge can be classified as either:

*   *Tacit*: knowledge stored in people's heads
*   *Explicit*: knowledge which has been written down or *codified*.

Explicit knowledge is the more familiar form of knowledge and is found in books, manuals and reports. Tacit knowledge is a much higher value form of knowledge, because we always know more than we can say (Sveiby, 1997). To be able to apply explicit knowledge to make decisions or take action, it must be made tacit. For example, you must read and understand a book in order to be able to apply the knowledge it contains.

Evidence-Based Medicine

Access to the latest medical research knowledge can mean the difference between life and death, an accurate or erroneous diagnosis, early intervention or a prolonged and costly stay in hospital (Ayres & Clinton, 1997). However research findings take a long time to filter into medical practice (Phillips, 1998). Empirical studies have shown that on average, there is an 8-13 year time lag (depending on the specialty) between a treatment being proven to work and its adoption in common practice. It has also been found that 70% of treatments currently in use do not have sufficient evidence to support that they are any more effective than doing nothing (Chalmers, 1993).

One of the major barriers to the implementation of research findings is the volume and geometric growth of the medical literature. It is not humanly possible to keep up with all the advances in all areas of medical research (Jordens et al, 1998). It is also difficult for medical practitioners to make sense of the often conflicting research findings in a particular area. Recognition of such problems led to the discipline of *evidence-based medicine* (EBM). Evidence-based medicine is a discipline which synthesizes research findings on the effectiveness of medical treatments to support clinical decision-making (Sackett et al., 1997). The aim of EBM is to bring research and practice closer together and reduce the time lag between the development of clinically proven treatments and their use in everyday medical practice.

One of the major methodological tools in EBM is the *systematic review* (Cochrane, 1972). Systematic reviews begin with an exhaustive search for published and unpublished research studies addressing a particular clinical issue (e.g., treatment of asthma). The next step is to critically evaluate the studies to identify which are of sufficient quality to contribute to decision making. The final step is to pool the results of the studies to arrive at a quantitative estimate of the effectiveness of the treatment(s). Synthesizing the research evidence is only the starting point for using research to improve practice. Equally important is the *dissemination* and *use* of this information. To make a practical difference, systematic reviews must be readily available to medical practitioners and must be actively used in everyday clinical practice. Reviews must also be regularly updated to take account of new research developments.

Evidence-based medicine is an application of knowledge management in the medical field although it predates the knowledge management literature by more than two decades. It focuses on synthesizing *explicit knowledge* in the form of research findings and using this knowledge in clinical decision-making.

Objectives of This Research

This paper describes a successful knowledge management project in one of Australia's state health departments. The objective of the project was to provide medical staff with on-line access to the latest medical knowledge at the point of care in order to improve the quality of clinical decision making. We believe this represents an important case study from both a theoretical and practical viewpoint:

- Examples of such unqualified success in IT projects are rare, and the reasons why it was so successful are worthy of investigation. In this sense, it represents a "unique and revelatory" case (Yin, 1994).
- Making the latest medical research available to clinicians is a critical issue in improving the quality of healthcare (Sackett et al., 1997). It therefore represents an important case in terms of its social significance.
- While much has been written on knowledge management, there is a lack of detailed knowledge about how to successfully conduct such projects. A major objective of this research was to understand *why* the project was so successful in order to provide guidance to practitioners wishing to conduct similar projects.
- The field of knowledge management is relatively new, and so far there have been few detailed case studies of knowledge management projects in practice. In addition, almost all previous empirical studies have been carried out in the private sector, even though the public sector comprises up to 65% of the economy in some industrialized countries (about 40% in Australia). This is one of the first published case studies of a knowledge management initiative in the public sector.

Outline of the Paper

The paper is organized in the following sections:
- Section 2 describes how the system was developed and implemented.
- Section 3 describes the content of the system, its availability, and how it is used in clinical practice.
- Section 4 describes the organizational impact of the system.
- Section 5 discusses the theoretical and practical implications of this research and draws some wider implications.

DEVELOPMENT AND IMPLEMENTATION APPROACH

Organizational Background

The organization in which the project took place was one of Australia's state health departments. This is one of the largest organizations in Australia in either the public or private sector, with a budget of over $6 billion in 1998 and over 100,000 staff. It is divided into a number of area health services, each of which administers healthcare facilities in a defined geographical area. Healthcare facilities include hospitals, community health centres, general practice, ambulance and other specialist facilities (e.g., drug and alcohol centres). Each area health service operates relatively autonomously and is managed by a chief executive officer, and is governed by a community-based area health service board.

Initiation of the Project

Historically, clinical information needs had not been well supported by investments in information technology. The majority of information systems in the organization support administrative processes (e.g., financial systems, payroll systems, patient administration systems), with very few systems directly supporting patient care. To address this issue, the Clinical Systems Reference Group (CSRG) was formed in September 1995 to specify requirements for clinical information at the point of care. The CSRG consisted of 50 clinicians from all health disciplines including hospitals, general practice, community health and universities. Each member of the group formed a network of clinicians within their own area health service to ensure the widest possible consultation. A report was produced in August 1996 and circulated widely throughout the organization.

In this report, the CSRG identified the need for access to the latest medical knowledge to support the objectives of evidence-based medicine (EBM). A proposal was developed for a project called the Clinical Information Access Project (CIAP) to meet this need, which was endorsed by senior management in December 1996. The stated objective of the project was, "To provide clinicians with access to on-line medical information to support clinical practice, education and research at the point of care."

A total of $800,000 was committed to purchase software licences, a Web server and the services of a project officer. It was recognized that rural and remote health facilities needed assistance with their infrastructure, so half of this funding was allocated for the purchase of PCs and Internet access.

Development Process

The development of the system took place in three phases:
- *Knowledge identification:* A pre-implementation survey of 2,757 clinicians and medical librarians was carried out to identify the most important information sources required to support clinical practice, education and research. The requirements identified included a combination of internal and external sources of information.
- *Knowledge sourcing:* Unlike most knowledge management projects, this project was primarily focused on external rather than internal

sources of knowledge (primarily medical research). There is a wide range of knowledge content providers and knowledge repositories (e.g., medical reference databases) in the medical field. The major part of this phase involved identifying the best possible sources of knowledge and negotiating licence agreements with them. For internal sources of knowledge, this involved converting documents to Web-based form.

- *Knowledge delivery:* A Web-based front end was developed to provide access to the information sources and to allow sharing of internal knowledge via a variety of mechanisms (bulletin boards, list servers, etc.).

The CIAP system went live on July 4, 1997, taking just over six months from its initial inception to implementation. The system has now been in operation for over three years and is still evolving. All of the top 10 priorities originally identified in the pre-implementation survey have been provided by the system.

Technology Architecture

The system operates using a single Web server located at head office. Maintaining the information in one location means that information is as up-to-date as possible. The Internet was chosen instead of intranet technology in order to maximize the "reach" of the system. Because of the geographic spread of the organization and the lack of a communications network linking all healthcare facilities, an intranet would have excluded a large proportion of clinicians. Using the World Wide Web, all that is required to use the system is a PC, a modem and Internet access. An added advantage of using the Internet is that it provides the flexibility for clinicians to access the system from home.

Implementation Process

Having developed the system, the next step was to get clinicians to use it. A series of presentations was given in each area health service to publicize the system and provide initial training. A CIAP representative, usually a practising clinician, was established in each area health service as a point of contact for:

- Publicizing and marketing the system
- Providing users with access to the system (via a common user name and password)
- Training and education
- Feedback from users of the system.

The CIAP representatives acted as local "champions" for the system. The organization has a history of IT initiatives that had failed as a result of local resistance to state-wide initiatives. The organizational culture is strongly decentralized and attempts to impose state-wide solutions on the largely autonomous area health services are strongly resisted. However although CIAP was a centralized project with a central development team, the implementation approach and organizational infrastructure are strongly area health service based, which fits with the organizational culture.

New information sources are continually being added to the system in response to requests by clinicians through their local CIAP representatives. As an indicator of how successful the system has been and how word of its success has spread, a number of health departments in other states have expressed interest in using the system. To support use of the system by other organizations, a consortium has now been established to allow other organizations to join the project. Five out of the seven Australian states are now members of this consortium and are in the process of implementing their own state-based versions of the system.

FUNCTIONAL CHARACTERISTICS

This section describes the facilities provided by the system and how it is used in clinical practice.

Knowledge Content

The knowledge content of the system was based on the requirements identified by clinicians in the pre-implementation survey. The system consists of three major components, each of which represents a different type of knowledge, as shown in Table 1 below:

- *Internal/External*: whether the knowledge originates from within the organization or external to the organization
- *Explicit/Tacit*: whether the knowledge is written down or in people's heads.

1. Medical Reference Databases (External, Explicit)

These are on-line databases which provide access to the latest medical knowledge and research. The databases include:

- Cochrane Library: Evidence-based medicine (EBM) reviews produced by the Cochrane Collaboration. The Cochrane Collaboration is an international not-for-profit organization which publishes systematic reviews in textbooks and electronic databases.

Table 1: Knowledge components

	INTERNAL	**EXTERNAL**
EXPLICIT	Clinical Policies and Protocols	Medical Reference Databases
TACIT	Listservers	

- MEDLINE: Produced by the U.S. National Library of Medicine, the MEDLINE database is widely recognized as the premiere bibliographic source of medical research. MEDLINE provides on-line searching with links to full-text medical journals.
- CINAHL (Cumulative Index to Nursing & Allied Health): Provides on-line searching capabilities with links to full-text nursing journals.
- PsychINFO: On-line searching capabilities with links to full-text psychiatry and psychology journals.
- Healthstar: On-line searching capabilities with links to full-text health administration journals.
- MIMS: This is Australia's most comprehensive pharmaceutical database and includes details of all known drugs, side effects, interactions and recommended dosages.
- Antibiotic Guidelines: For decision support in prescribing antibiotics (antibiotics are generally poorly prescribed and oversubscribed in practice).
- Micromedex: Poisons and toxins information.
- Full-text journals and textbooks: A range of full text journals and medical textbooks are also available on-line.

Together these reference databases represent the most authoritative sources of medical research knowledge available. Access to primary research sources (e.g., MEDLINE, full-text journals) are provided in addition to EBM reviews, as such reviews are currently only available for a small percentage of medical conditions. All of these reference databases have been developed and maintained by third-party information providers, so represent sources of *external knowledge.*

2. Clinical Policies and Protocols (Internal, Explicit)

Clinical policies and protocols play a critical role in medical practice (Wilson et al., 1995; 1999) and represent one of the most important sources of explicit knowledge in the organization. Each hospital defines its own policies and protocols for handling particular types of cases (e.g., cardiac arrest, road trauma). CIAP allows clinicians to post clinical policies and protocols for peer review. The purpose of this is to encourage collaboration and sharing of knowledge and to facilitate development of best practice policies and protocols. Documents are posted in the following categories:

- Policies
- Procedures
- Protocols
- Clinical pathways
- Clinical practice guidelines.

3. List servers (Internal, Tacit)

List servers are provided for on-line discussion of problems and issues in particular areas of interest, for example, asthma, stroke and ethical issues. A list server is an electronic mailing list which automatically distributes e-mail messages to everyone listed as a "subscriber." This facilitates the exchange of ideas and experiences among clinicians across the state. This is especially useful for clinicians in rural and remote areas, who have less opportunity for face-to-face exchange of knowledge. List servers facilitate the transfer of *tacit knowledge* between individual clinicians.

System Availability

24-Hour Access

A major advantage of the system is that it is available 24 hours a day, 7 days a week to all clinicians across the state. Access to the system is provided via a common user name and password assigned to each hospital, available on request from CIAP representatives. Prior to CIAP, a subset of the information provided (e.g., medical journals, MEDLINE) was available through medical libraries attached to the larger hospitals. However medical libraries are generally not open after-hours or on weekends and many smaller hospitals do not have libraries at all. As well as this, in a "life and death" situation, clinicians do not have the time to go to the library.

Figure 1 shows system usage by hour of day; the peak volume of almost 25,000 hits per hour shows how popular the system has been. The high usage figures outside of normal business hours partly reflects the fact that health is a 24-hour profession, but also that many clinicians don't have time to access information during working hours. For example, "as a clinical nurse consultant with an average of a 10-hour day, the luxury of being to be able to access this information from home has been invaluable to myself, colleagues and patients. I find it difficult to get to the library during work hours."

Access at the Point of Care

Clinicians need to make life or death decisions in the workplace every day. In an emergency situation, they do not have the time to go to the library to find out the best treatment for a particular condition or the possible side effects of a drug. Having such information available at the point of care can make the difference between life and death for a patient.

Access from Home

Use of the Internet enables the flexibility for clinicians to access the system from home. This enables them to use the system for educational purposes...to upgrade their qualifications and maintain accreditation. The usage statistics show that 50% of system usage is from home (Figure 2).

The most likely reasons for the high usage levels from home are:
•   Clinicians often do not have access to PCs and the Internet in the workplace. The lack of IT infrastructure in the clinical workplace is by far the greatest limitation on the usage of the system in everyday clinical practice.
•   Clinicians often do not have time to use the system during working hours (as mentioned previously).

Figure 1: System usage by hour of day (April 1999)

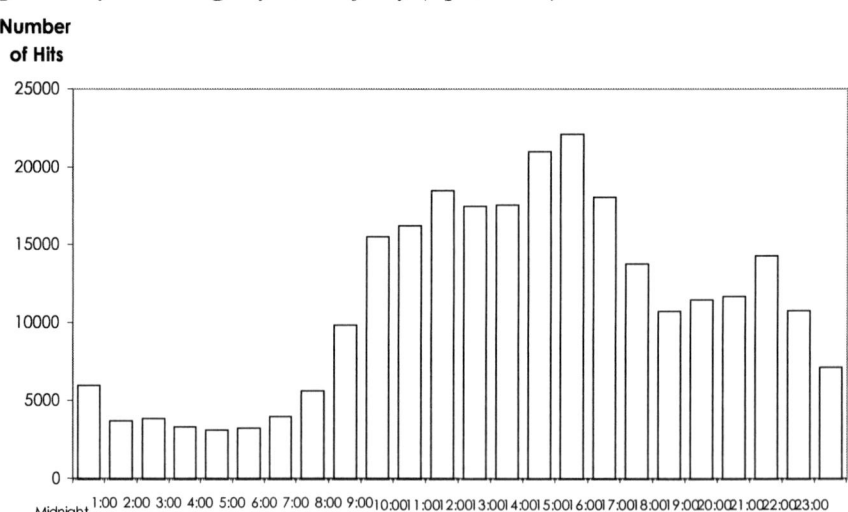

Hour of Day

Access from Rural and Remote Locations

Clinicians in rural and remote areas frequently don't have access to medical libraries and have to travel long distances to get access to them. As a result, CIAP has received a particularly positive response from rural clinicians. CIAP has been successful in reducing information inequalities between clinicians in different parts of the state. Some of the survey responses received were:

*   "Reduces isolation of rural clinicians."
*   "I am an emergency physician and also DCT in a rural hospital. Access at the point of care to current information is a necessity we were previously denied. The Web site is brilliant and I am especially delighted to see Micromedex as this is a superb package for emergency departments but usually too expensive for smaller departments (where it is often most useful)."

Scenario–How the System Is Used

To illustrate how the system may be used in everyday clinical practice, consider the case of a patient who presents at a hospital with salmonella. The RMO (registered medical officer) decides to treat the case with antibiotics and consults the system to determine the most effective one to use. Antibiotics are commonly recommended early in the illness and are widely used...80-90% of culture proven cases (Sirinavan & Garner, 1998). The rationale is that antibiotics will shorten the duration of diarrhoea and prevent the serious complications associated with the infection (Farthing et al., 1996).

From the opening screen of CIAP, the RMO enters the Cochrane Library and searches on salmonella. The search returns five reviews, one of which is a comparison of antibiotic treatments for salmonella. Clicking on the relevant item brings up a full-text version of the review (Figure 3).

Figure 2: Access location

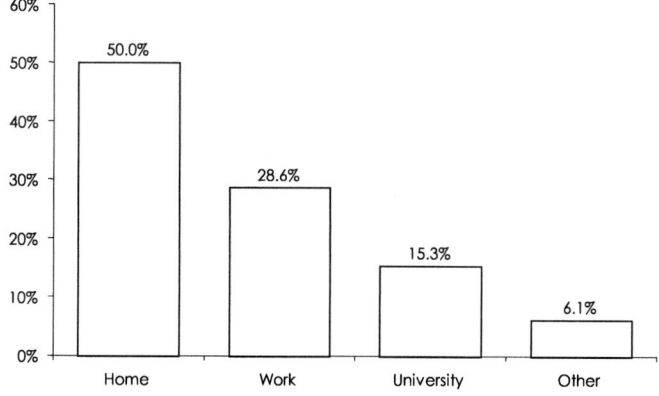

The conclusions from the study were: Fifteen trials on 857 participants, in which 366 were infants and children, were included. The investigated drugs were neomycin, chloramphenicol, ampicillin, amoxycillin, cotrimoxazole, norfloxacin, fleroxacin, and ciprofloxacin. There appears to be no evidence of a clinical benefit of antibiotic therapy in otherwise healthy children and adults with non-severe salmonella diarrhoea. Antibiotics appear to increase adverse effects and they also tend to prolong salmonella detection in stools. (Sirinavin & Garner, 1998).

Based on this information, the RMO should not prescribe antibiotics to the patient. This will reduce the costs of treatment and also reduce the chances of adverse drug effects. This example emphasizes the fact that many treatments routinely used in practice have no effect (or in this case, a negative effect) on patient recovery, thus adding to healthcare costs without improving health outcomes. It is also an example of the misuse of antibiotics in common practice.

The time taken from the CIAP opening screen to access the Cochrane Library, perform the search, look up the full-text article and extract the required information was 2 minutes and 12 seconds, which is well within the time frames required for clinical practice.

ORGANIZATIONAL IMPACT

This section describes the benefits of the system and its impact on clinical practice.

Figure 3: Full-text review

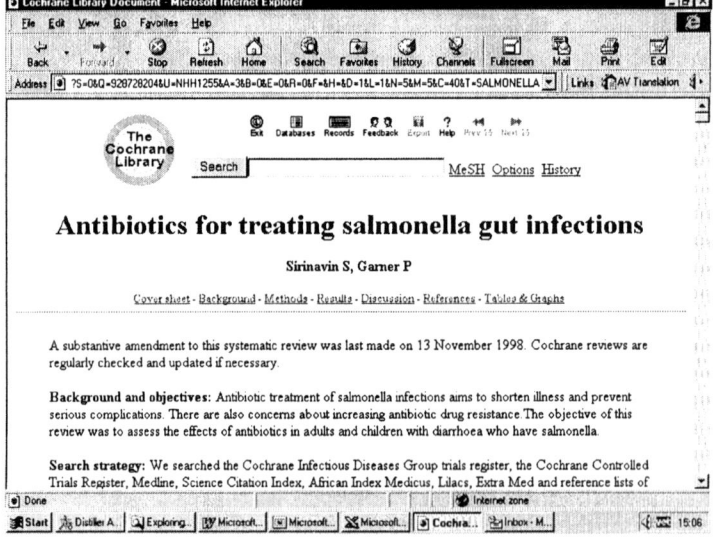

System Usage

In terms of usage, the system has shown a spectacular and sustained increase in usage over its lifetime. The graph below (Figure 4) shows an exponential increase in information usage since the beginning of 1998. Usage of information is one of the most frequently reported measures of information system success, but is only relevant when use of the system is voluntary (De Lone & McLean, 1992). It is clearly appropriate in this case, as clinicians are not forced to use the system and have to request access to it.

User Satisfaction

User satisfaction is probably the most widely used measure of information system success. One reason for this is that it has a high level of face validity; it is hard to deny the success of a system that people like (De Lone & McLean, 1992). Reaction to the system has been almost universally positive, which has surprised even the project team. A state-wide survey carried out in 1998 (the Information Needs Analysis Survey) found that CIAP had the highest levels of user approval of any system in the organization (Moody et al., 1998). The only negative responses received about the system were from people unable to get access to it. The responses to the on-line survey were also very positive. This is unusual so soon after a system has been implemented, when complaints usually outnumber compliments. The only complaints received were about speed of access, which is a known limitation of using the Internet as a delivery mechanism.

Figure 4: System usage statistics: January 1998-August 1999

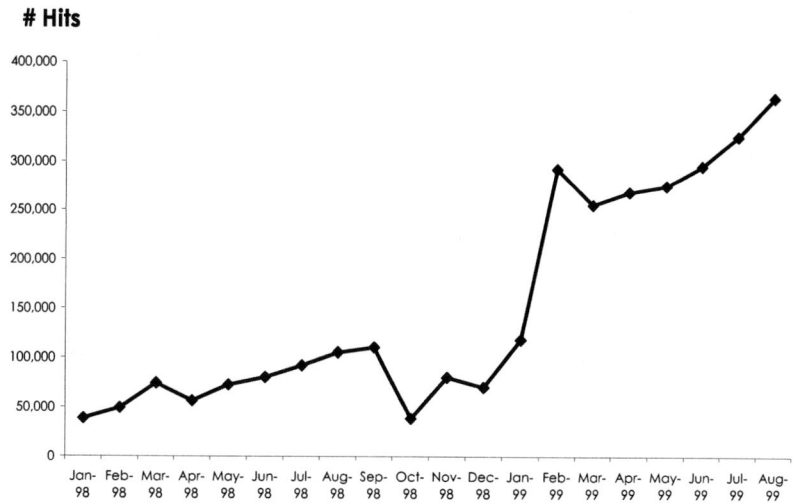

Impact on Patient Care

Organizational impact measures improvements in organizational performance as a result of the project, either in terms of efficiency or effectiveness gains (De Lone & McLean, 1992). This is probably the most important success indicator of all, in that it measures the real benefits of the project for the organization. The objective of this project was to improve the quality of clinical decision-making and thereby improve the quality of patient care. Both anecdotal evidence and perceptions of clinicians support the conclusion that it has been successful in doing this. An on-line survey of clinicians using the system revealed that 90.82% felt that it had improved patient care.

There have also been a number of reported cases where access to CIAP has actually saved lives. For example:

- In the North West Area Health Service, a clinician was able to save a patient in a critical condition suffering from the Lyssavirus. The Lyssavirus, which is acquired from contact with bats, causes encephalitis in humans and can be fatal if not treated quickly. However because it is a relatively new disease (the first case was discovered in Australia in May 1996), it does not yet appear in medical textbooks. Using MEDLINE, the clinician was able to look up and apply the appropriate treatment.
- In the Far West Area Health Service, a clinician was able to save a patient who was admitted in a critical condition in the middle of the night, suffering from meningitis. The patient did not respond to the normal treatment for this condition, indicated by the standard clinical protocol. A MEDLINE search revealed a new drug treatment which had immediate results.
- In the North Coast Area Health Service, a clinician was able to save a child who had been bitten by a funnel Web spider by looking up the antivenom in the poisons database in Micromedex. For the anti-venom to be effective, it must be applied within 30 minutes, so having on-line access to this information meant the difference between life and death.

A more detailed empirical evaluation of the effect of the system on clinical practice is currently in progress as part of this research.

CONCLUSION

Summary of Findings

For medical research to make a practical difference and improve health outcomes, research results must be readily available to clinicians in the workplace and must be actively used and implemented in everyday clinical practice (Jordens et al., 1998; Phillips, 1998). This paper has described a system which makes the latest medical research available on-line to support clinical decision-making at the point of care. Having such information available at the point of care may make the difference between life and death for a patient.

The system was implemented in a very short time, with a relatively low budget and using very simple technology, yet has had quite a dramatic organizational impact. It represents a world-class example of how knowledge, delivered using the Internet, can be used to support medical practice. It is a rare example of an information system which has the potential to save lives and, in a number of documented cases, actually has. On the other hand, there are a number of technical, organizational and financial barriers to the widespread adoption of the system in clinical practice which need to be overcome.

Theoretical and Practical Significance

A major limitation of previous empirical research in this area is that it has taken place almost exclusively in the private sector. This is one of the first empirical studies of a knowledge management project in the public sector.

We have provided a description of a successful knowledge management project that will help practitioners better understand why such projects are successful. The results of this study can be easily generalized to other similar organizations, in particular, to health departments in other states and other countries. A similar system could be successfully implemented in any large health department following the principles described here. The results could also be generalized to healthcare providers in the private sector, for example, private hospitals and general practice where the principles of evidence-based practice are equally applicable.

Wider Implications

This system has the potential to revolutionize the practice of healthcare. It provides a mechanism for moving away from experience/anecdotally based medicine and towards evidence-based practice. While the project has been very successful so far, its impact has still been relatively small—it has been limited to the public health system, and mainly senior clinicians within hospitals. Its penetration into clinical practice has been limited by the lack of technology in the workplace, restrictions on access and financial constraints. If it was made available to all clinicians in the organization, its impact would be increased significantly. If access was extended to private practice, particularly GPs, who form the "front line" of the health system, the impact on the quality of healthcare would be enormous.

If use of the system was extended to consumers, it could revolutionize the healthcare system. For example, if you are a patient being treated for a particular medical condition, you could look up the latest medical evidence about the treatment you are being given. If you find that the treatment does more good than harm, you would rightly question your doctor's decision. This would give an enormous amount of power to consumers (and take an equivalent amount away from clinicians). However there would likely be considerable resistance to such a move from the medical profession, as it could open up clinicians to questioning of their decisions and possible litigation.

The medical profession currently enjoys almost "god-like" status in the community, largely as a result of the knowledge differential between medical practitioners and consumers. Patients rarely question their doctor's advice and there is very little accountability for their actions. It is one of the few professions which has maintained its status and has remained resistant to external scrutiny or questioning of its practices. Making the latest research evidence publicly available would create a race of "intelligent consumers" and would remove the large knowledge gulf that currently exists between doctors and patients.

ENDNOTE

[1] An earlier version of this paper was presented at the International Conference on Decision Support Systems (DSS'2000), Stockholm, Sweden, July 9-11, 2000.

REFERENCES

Ayres, D. H. M. and Clinton, S. (1997). The user connection: Making the clinical information systems vision work in NSW Health. *Health Informatics Conference (HIC '97)*, Sydney.

Broadbent, M. (1997). The emerging phenomena of knowledge management, *The Australian Library Journal*, February, 6-25

Chalmers, I. (1993). The Cochrane collaboration: Preparing, maintaining and disseminating systematic reviews of the effects of health care. In Warren, K. S. and Mosteller, F. (Eds.), *Doing More Good Than Harm: The Evaluation of Health Care Interventions, Annals of the New York Academy of Sciences*.

Cochrane, A. L. (1972). *Effectiveness and Efficiency: Random Reflections on Health Services*, London: Royal Society of Medicine Press.

Davenport, T. H., De Long, D. W. and Beers, M. C. (1998). Successful knowledge management projects, *Sloan Management Review*, 39(2), 43-52.

De Lone, W. H. and Mclean, E. R. (1992). Information systems success: The quest for the dependent variable. *Information Systems Research,* 3(1).

Farthing, M., Feldman, R., Finch, R., Fox, R., Leen, C., Mandal, B., Moss, P., Nathwani, D., Nye, F., Ritchie, L., Todd, W. and Wood, M. (1996). The management of infective gastroenteritis in adults. A consensus statement by an expert panel convened by the British Society for the study of infection. *Journal of Infectious Diseases*, 33, 143-152.

Hansen, M., Nohria, H. and Tierney, T. (1999). What's your strategy for managing knowledge? *Harvard Business Review*, March/April.

Jordens, C. F. C., Hawe, P., Irwig, L. M., Henderson-Smart, D. J., Ryan, M., Donoghue, D. A., Gabb, R. G. and Fraser, I. S. (1998). Use of systematic review of randomized trials by Australian neonatologists and obstetricians, *Medical Journal of Australia*, March 16.

Martin, W. (1999). Knowledge: Changing the focus from production to management (Working Paper), *RMIT Business*, Melbourne.

Moody, D. L, Anderson, L. F., Blackwell, J., Johnston, K. and Barrett, K. (1998). Measuring outcomes of information management—Lessons from the health industry. *3rd Australian DAMA Conference*, Old Parliament House, Canberra, October 12-13.

Nonaka, I. and Takeuchi, H. (1995). *The Knowledge Creating Company*, Oxford University Press, New York.

Phillips, P. A. (1998). Disseminating and applying the best evidence, *Medical Journal of Australia*, March 16.

Sackett, D. L, Richardson, W. S, Rosenberg, W. and Haynes, R. B. (1997). *Evidence Based Medicine: How to Practice and Teach EBM*, New York, Churchill Livingstone.

Sirinavin, S. and Garner, P. (1998). Antibiotics for treating salmonella subinfections. (Cochrane Rutter). Abstract. The Cochrane Library, 2, 2001. Retrieved July 25, 2001 from the World Wide Web: http://www.cochrane.de/cc/cochrane/revabstr/ab001167.htm.

Sveiby, K. E. (1997). *The New Organizational Wealth: Managing and Measuring Knowledge-Based Assets*, Berret-Koehler Publishers, San Francisco.

Weed, L. L. (1997). New connections between medical knowledge and patient care, *British Medical Journal*, 315, July 26.

Wilson, R. McL., Harrison, B. T, Gibberd, R. W. and Hamilton, J. D. (1999): An analysis of the causes of adverse events from the quality in Australian health care study, *The Medical Journal of Australia*, May 3.

Wilson, R. McL., Runciman, W. B, Gibberd, R. W., Harrison, B. T, Newby, L. and Hamilton, J. D. (1995). The quality in Australian health care study, *The Medical Journal of Australia*, November 6.

Yin, R. K. (1994). *Case Study Research: Design and Methods* (2nd ed.), Sage Publications, Thousand Oaks.

Chapter XII

Structured Content and Connectivity of Medical Information–The Medical Data Web

Walter Fierz
Institute for Clinical Microbiology and Immunology, Switzerland

INTRODUCTION

To a large extent, medicine is an information-processing endeavor. It is all the more surprising, therefore, that computer technology has not yet developed very far in successfully supporting this activity. The areas where computer support is most advanced in medicine are signal processing and 'data'[1] generation, as well as in the administrative domain. However, the huge area that encompasses the main part of the physician's clinical activity has only marginally been touched upon by information-processing systems. The reason for this might well lie with the problem that medical 'information' is mainly a matter of connecting pieces of data together. The sheer complexity of these connections is, however, beyond classical database technology. The best and easiest way to express such complex information has been written (and spoken) text, until now. A long tradition of how written medical reports are formulated and stored in patient records and how scientific reports are published in medical journals has been built up.

When one analyses such reports and medical thinking in general, it becomes very obvious that data elements, like results of laboratory tests, X-ray pictures, or any other clinical observations carry little information on their own. They become meaningful only when they are embedded in a specific context and connected to other data elements. It is here, where computer technology is badly needed not to replace the doctors' thinking but to support their decision process in a way that the patient can profit from this connectivity and from global medical knowledge.

Connectivity of Medical Information

A typical medical data element is a result from a laboratory test. To properly interpret such a result, connections have to be made to reference values, to test parameters like method, sensitivity, specificity, and likelihood ratio, to preexisting results of the same test for that patient, to other test results, to the clinical context, and last but not least to the scientific literature. Such connections might be precise and stringent, or they might be loose and interpretable. There might be lists involved from which the doctor has to choose the appropriate item, or the required target element might be not accessible or be even missing. The connection might be to targets within the same domain of activity, or they might point to distant locations like other databases or other documents, in the same information system or some other system. In any case, medial data connectivity is not of a quality that can be adequately modeled in a relational database.

To give an example, we could look at CD4-lymphocytes, a laboratory parameter that is regularly used to monitor HIV infection (Figure 1). A number of 250 for CD4 would probably give the treating physician enough information because the context is obvious to her. But the connectivity is extensive (see figure): She assumes that the number 250 refers to cells per microliter in the metric system, and CD4 refers to a subpopulation of lymphocytes, a type of white blood cell. She assumes that the measurement has been done by a method known as flow cytometry in a sample collected from the peripheral blood in an EDTA tube. It will not be obvious what reagent has been used for staining the CD4-cells, what type of apparatus to measure them, and what confidence interval such a measurement has. She knows that the reference value in the normal population is above 500, but the link to the publication that asserted that value and described that control population is not apparent. Of course she has to know from which patient the sample has been drawn and at what time and date. But then she should also know earlier results of the same test to compare to the current result. CD4 values are not only expressed in absolute cell numbers per microliter but also relatively as percentage of

Figure 1: The immediate connectivity of a laboratory value: Laboratory data, such as a value of 250 for CD4 cells per microliter of blood, or any other clinical observation is only informative when it is embedded in a connected context. The physician provides this connectivity implicitly with his knowledge and experience and expresses the information in free text. Other than that, an electronic patient record needs to explicitly express the connectivity if the information is to be processed by machines, in order to support the physician in his clinical decisions.

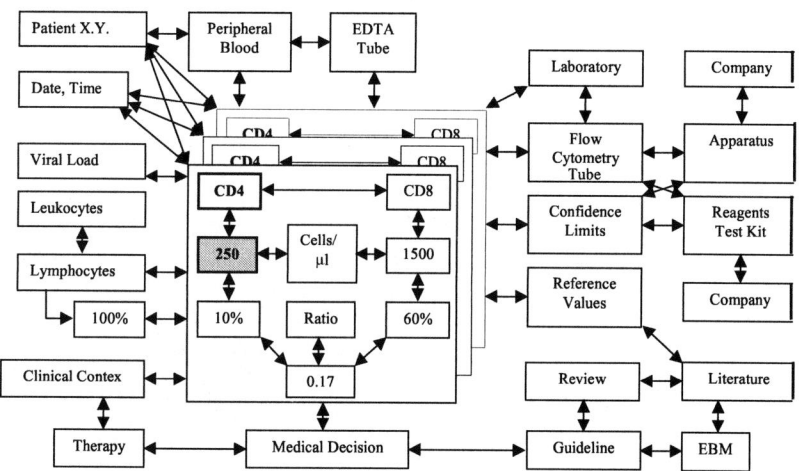

The roles and types of links (←→) are not indicated.

CD4 cells among all lymphocytes. CD4 values have to be related to CD8 values, another lymphocyte subpopulation, and the ratio between CD4 and CD8 values is calculated. These results have again their own reference values. Another laboratory parameter, the viral load, moves in close relationship to the CD4 values. Furthermore, these laboratory parameters are embedded in a wider context of other clinical information. There is quite some volume of scientific literature about the impact of CD4 values and viral load on treatment decisions. This literature has been reviewed, meta-analyses for evidence-based medicine have been published, and guidelines have been formulated and adapted to the local use.

For a machine to support the doctor in decision making, none of these connections are "obvious." All of them have to be expressed explicitly in some kind of model. Of course, a laboratory test result is just a representative example and other clinical data have the same high degree of essential connectivity. The challenge is to be aware of these connections and to exploit them productively.

Structures in Medical Information

More rigorous connections can be expressed in structures. When one reads a medical record, a lab report, or a scientific paper, one always finds structures, to varying degrees. A medical record might contain categories like "subjective," "objective," "assessment," and "plan"; a scientific paper usually has chapters like "summary," "introduction," "material and methods," "results," and "discussion." Both types of documents might contain lists, tables, indices, table of contents, glossaries, etc. However, the degree of 'granularity' of these explicit structures is usually low. When one looks at the blobs of content within them, one finds further implicit structures buried within text, tables, figures, etc. However, these structures are only available to the informed reader and not directly to machine processing. To be sure, text-analyzing systems might be able to extract some structure from free text, but only the informed reader can fully analyze the implicit information structures, because making these structures explicit is an active, energy-consuming, value-adding, intelligent process. Such types of structuring ensure that the created structures are based on 'semantic' grounds rather than on mere syntactic rules like, for example, in an alphabetic index.

REQUIREMENTS FOR A MEDICAL INFORMATION SYSTEM OF STRUCTURED AND CONNECTED DATA

The nature of structured and connected medical information poses some particular requirements on an information system in order to enable it to process not only the data themselves but also that part of information that resides in the connectivity and structure. These requirements are summarized in the following five points.

1. Granularity of Data Elements

Ideally, each atomic data element that cannot be divided any further should be individually accessible in order to process it with computer programs. However, not every application needs full granularity, and it is a matter of balancing the needs of the application against the practicability and cost of granularity. For an information system to support various needs of applications, it is advantageous to support hierarchical structures that allow an unfolding of the granularity according to actual needs.

2. Semantics of Data, Links and Structures

There should be a way to attach some meaning to individual data elements. However, more than that, links between data elements also have meaning, as well as the structures that contain the data elements. All this semantic information should be represented in a standardized way.

3. Database for Structured and Linked Information

Of course, to be long-lived, structural information has to be stored somehow, together with the data elements. Ideally, such a storage system is independent from previous and future application systems, i.e., it should use a standardized technology. This is very important, since medical information might be used throughout a patient's lifetime and present applications cannot foresee future applications. The most simple, flexible, and generally agreed-upon standard will provide the highest chance of long survival of the structures, links, and data.

4. Query System

A query system is needed that is able to parse the information and navigate the various structures and links in order to extract the necessary information, again in the form of structures, links, and data elements. Such a query system should be easy to handle and ideally uses natural language and graphical display. The query process should be interactive and flexible.

5. Interactive, Adaptable User Interface

The user interface of a medical information system has to be able to display that part of information that is intrinsically contained in the structure and connectivity of the data. Such display should be adaptable to a given doctor's role and way of thinking. The user should be invited to interact with the system and to extract that piece of information and those connections that are most pertinent and useful for his decision process. On the other hand, the interface should remind the user of important facts that might have escaped her attention. What's more, it should allow associative browsing that gives serendipitous discovery of relevant information a chance.

Information extracted from such an interactive process, as well as information gained in this process, should be stored again for future use. The stored information should include the user-added connectivity and structure.

THE SOLUTION

To solve these problems of structuring and connecting medical data in a machine-processable way, the emerging 'Web' technology seems to provide the appropriate concepts and methods. Two major paradigms are fundamental to both the current success of the Web and the future of a 'Medical Data Web.' First, the 'hypertext' concept is instrumental in solving the problem of connectivity, and second, 'markup' languages provide the technology for structuring data. The two paradigms will be introduced and discussed as to their aptness for enabling the Medical Data Web in the following two sections.

The Hypertext Paradigm

"Everything is deeply interwingled." (Ted Nelson)

In July 1945, Vannevar Bush, President Roosevelt's science advisor during World War II, outlined in "As We May Think," an article in *The Atlantic Monthly,* the ideas for a machine called *Memex. Memex* would have the capacity to store textual and graphical information in such a way that any piece of information could be arbitrarily linked to any other piece. He wrote:

Our ineptitude in getting at the record is largely caused by the artificiality of systems of indexing. When data of any sort are placed in storage, they are filed alphabetically or numerically, and information is found (when it is) by tracing it down from subclass to subclass. ... The human mind does not work that way. It operates by association. With one item in its grasp, it snaps instantly to the next that is suggested by the association of thoughts, in accordance with some intricate Web of trails carried by the cells of the brain.

Twenty years later, in 1965, Ted Nelson coined the terms *hypertext* and *hypermedia*. After another 20 years, in 1985, Xerox released *NoteCards*, a LISP-based hypertext system, and in 1987 Apple Computers introduced *HyperCard*, which popularized the hypertext model.

Hypertext is the presentation of information as a linked network of nodes that can be navigated freely. The granularity of the link destinations and the flexibility of how the links can be traversed are two important properties that qualify hypertext systems. These properties are also decisive in a Medical Data Web. Although the hypertext model was originally designed for textual and graphical nodes, it may also be used for data in general. In this way, it may well be a good fit for the requirements posed by the connectivity of medical data. However, the hypertext model does not make any assumptions about the structured context of the linked elements. Therefore, hypertext alone is not sufficient as a model for a Medical Data Web.

The World Wide Web

In the nineties, the hypertext paradigm made its global appearance in form of the 'World Wide Web' (WWW). In 1989, Tim Berners-Lee, the founder of the WWW, wrote in his proposal to CERN about information management:

> *The problems of information loss may be particularly acute at CERN, but in this case (as in certain others), CERN is a model in miniature of the rest of world in a few years time. CERN meets now some problems which the rest of the world will have to face soon.*

However, even at time of the above statement, the medical domain was already suffering from the same problem of data loss. Nevertheless, the successful implementation of Tim Berners-Lee's proposal and its global acceptance in form of the WWW has made the need for a solution in the medical domain even more obvious. To be sure, the WWW has already found a place in the medical profession, and many doctors already use it in their daily work. Scientific medical publication is gradually shifting towards on-line publication and hypertext links are becoming more and more popular in scientific writing. However, the granularity of the links is usually still very coarse, since the linked nodes are mostly entire scientific articles.

However, when it comes to storing and handling medical patient data, the hypertext paradigm has not yet been adopted. There, if anything, the classical relational database model is still preferred, and it is predominantly used for numerical, statistical data and for administrative purposes. One reason might well be that a suitable model for highly linked and structured data has, so far, been missing.

The details of how the hypertext concept is rolled out with modern Web technology is discussed in the following paragraphs in context with the second paradigm, the markup language of the Web.

The Markup Paradigm

The implementation of the hypertext concept strongly depends on the availability of a technology that allows it to address the granularity of the nodes that are to be linked. Ideally, the method of structuring does not reflect a mere syntactic approach like, for instance, alphabetic indexing or numbering of documents, paragraphs, figures, etc., but provides a way to attach some meaning to the data elements. A doctor might like to find, for example, links to reference values for CD4 cell numbers measured by flow cytometry, rather than a price list of CD4 reagents used for measuring these cells. Fortunately, in parallel with the development of the hypertext concept, the second ingredient for a Medical Data Web has emerged: the markup concept.

SGML/XML–The Enabling Standards

Markups are defined as machine-manageable annotations within a multimedia document. Originally, they were used in the printing domain to attach formatting information to identified sections of text. More generally, markups can be used to annotate any type of data with some semantic information. A particularly interesting application in this context is the use of markups to uniquely identify data elements and to semantically type the data. In this way, it is possible to aggregate the data into a regular structure.

In the sixties, Goldfarb and others undertook a successful attempt to standardize the use of markups by creating the general markup language SGML (Standard Generalized Markup Language, ISO 8879:1986; Alschuler, 1998; Goldfarb, 1990). SGML provides a generic framework for markup constructs; the only major constraint is its restriction to a hierarchical structure of markup elements with a single root element.

Tim Berners-Lee's concept to use markup to implement a hypertext system was later to become the popularizer of SGML. His Hypertext Markup Language (HTML) could actually be defined as an application of SGML syntax. However, the restriction of the open semantics of SGML to formatting purposes of HTML has made the latter somewhat too rigid. But recently, a renaissance of SGML took place when the eXtensible Markup Language (XML) was born as a simplified but still powerful and open child of SGML, which is gaining very quickly in popularity now (Bray, 1998; Bray, Paoli, Sperberg-McQueen, & Maler, 2000). The development of this open and enabling framework of markup standards has been described elsewhere in more detail by Fierz and Grütter (2000). This chapter will concentrate on the use of SGML/XML as a solution for the Medical Data Web.

XML in Healthcare

Prior to the development of XML, the capabilities of SGML to structure data in a standardized way were recognized as promising for application to patient records. The advent of XML has encouraged these early visions and several attempts are now under way to make use of the markup paradigm for medical information. Two examples shall be mentioned below.

The Oswestry Project of the National Health Service (NHS) in the United Kingdom

The largest implementation of SGML/XML for medical records so far has occurred in the United Kingdom. A National Health System project with the aim of structuring patient records was originally based on SGML, but later moved on to XML. Over the last six years, over 300,000 patient record documents from the Robert Jones and Agnes Hunt Orthopaedic and District Hospital NHS Trust in Oswestry (Wales) were translated into XML (Leeming, Garnett, & Roberts, 1999). This is, to our knowledge, the first large-scale implementation of the markup paradigm for electronic patient records. Several lessons were learned, but one important conclusion from this project was:

> *The major obstacles preventing the adoption of EPR are cultural and political.*

This remark reflects the view that XML is more than just a new technology; XML leads to a new way of thinking.

Clinical Document Architecture Framework

Another important effort to employ XML in healthcare is being undertaken by a technical committee of the HL7 named *Structured Documents*. It proposes a 'Clinical Document Architecture' (CDA) as a markup standard that specifies the structure and semantics of clinical documents for the purpose of exchange. A draft for a CDA framework was published August 4, 2000 (Alschuler et al., 2000):

> *A CDA document is a defined and complete information object that can include text, images, sounds, and other multimedia content.*
>
> *The CDA is envisioned as an extensible and hierarchical set of document specifications that, in aggregate, define the semantics and structural constraints necessary for the representation of clinical documents.*
>
> *"Levels" within the architecture represent a quantum set of specializations, to which further constraints can be applied:*
>
> *• "CDA Level One" is the root of the hierarchy and is the most general document specification. RIM classes (Health Level Seven, 2000) are used in the specification of the document header, while the document body is largely structural, although terms from controlled vocabulary can be applied.*
>
> *• "CDA Level Two" will be a specialization of CDA Level One, and will constrain the set of allowable structures and semantics based on document type code.*
>
> *• "CDA Level Three" will be a specialization of CDA Level Two that*

*will specify the markup of clinical content to the extent that it can
be expressed in the HL7 RIM.*

The notion of an ever-finer granularity with each level of the CDA reflects well the requirement for a flexible granular structure of a patient record. The project also demonstrates how the markup technology is apt to assign semantics to clinical documents, in this case as defined by the HL7 Reference Information Model RIM (Health Level Seven, 2000).

In October 2000, HL7 approved the CDA Level One, which is expected to be published soon as an ANSI-approved standard.

XML Schemas and Mediation Between Schemas

Any attempt to specify some constraints to using markups for medical documents must face the problem of how to communicate the marked up information to a different system that uses other markup schemas. A physician from the Oswestry project of the NHS might like to exchange some data with a physician who works with the CDA of HL7. It is, therefore, important that XML provides the method of mediation between different XML schemas. For this purpose a language for transforming XML documents into other XML documents called XSLT has been recommended by the W3C (Clark, 1999). Furthermore, it will be important that particular 'document type definitions' (DTDs) and schemas are made public in schema repositories.

XML for Data and XQuery for Their Retrieval

Although SGML/XML is document-oriented and has its origins in the publishing industry, the specification can be used for any type of data. Since the original way of defining an XML document type with a DTD is somewhat restricted and does not allow for data typing, endeavors are under way by the W3C to specify a schema language. These include: XML Schema Part 1 specifies structures (Thompson, Beech, Maloney, & Mendelsohn, 2000) and Part 2 data types (Biron & Malhotra, 2000). This XML schema will be itself extensible, so any type of application-specific data can be typed. This is very important when XML is applied to medical data. It provides a means of representing semantic information of the data elements. Referring to our example in Figure 1, one could envisage defining a data type for blood lymphocyte quantities as cells per microliter.

For storage, XML data could either reside in a relational database, or they might be stored directly as native XML. The latter will probably become more and more important, because initial experience shows it to be better in performance. However, to be useful, XML storage needs a query language for data retrieval. Again, such a specification is under development by the W3C

(Chamberlin, Fankhauser, Marchiori, & Robie, 2001). An interesting proposal is Quilt, an XML query language for heterogeneous data sources (Robie, Chamberlin, & Florescu, 2000):

> *Quilt is able to express queries based on document structure and to produce query results that either preserve the original document structure or generate a new structure. Quilt can operate on flat structures, such as rows from relational databases, and generate hierarchies based on the information contained in these structures. It can also express queries based on parent/child relationships or document sequence, and can preserve these relationships or generate new ones in the output document. Although Quilt can be used to retrieve data from objects, relational databases, or other non-XML sources, this data must be expressed in an XML view, and Quilt relies solely on the structure of the XML view in its data model.* (Introduction)

Based on Quilt, a working draft of a query language for XML (XQuery) has been issued by the W3C on February 15, 2001 (Chamberlin, Florescu, Robie, Siméon, & Stefanescu, 2001).

Storing and retrieving typed data of any kind will be pivotal for the use of XML in Medical Data Webs.

Displaying XML–Stylesheets (XSL) and Scalable Vector Graphics (SVG)

Whereas HTML is predominantly used for formatting purposes in Web browsers, SGML/XML documents alone do not primarily contain any display information. Formatting information is defined in a separate specification called Extensible Stylesheet Language (XSL; Adler et al., 2000). This separation of content and style makes XML extremely well suited for flexible and user-adapted display. With this means, the same patient record content can be displayed, for example, on a desktop as well as on a portable computer device.

Since XML content is not restricted to text alone but can also be used for numerical or other data, the question soon arises how to display numerical content in a graphical way. The openness and flexibility of XML quickly led to a formulation of a graphical syntax to define the ingredients of 2D-vector graphics. The W3C has recently approved a candidate recommendation called 'Scalable Vector Graphics' (SVG) that was readily adopted by the industry (Ferraiolo et al., 2000). An SVG browser plug-in for Internet browsers is freely available from Adobe company (2000).

The important property of SVG is its dynamic nature. As such, graphic templates are used in Web pages and browser applications that enable them to dynamically and interactively display XML data. Since SVG is an XML application, it also provides the opportunity to include hypermedia links required by the connectivity of the Medical Data Web (see below). A graphical display of the human body, for example, could be dynamically adapted using underlying patient data to exhibit hot spots of clinical problems, and links to the corresponding details of data could be included in the graphics.

The Linking Problem–HyTime

One reason why the implementation of Tim Berners-Lee's HTML concept has been so successful is the combination of the markup paradigm with the hypertext paradigm. At the same time, one of the limitations of HTML is that the linking possibilities are rather restricted.

A much more general way of specifying links has been defined in an SGML-based ISO standard called HyTime (Hypermedia/Time-based Structuring Language, ISO/IEC 10744:1997; Goldfarb, Newcomb, Kimber, & Newcomb, 1997). HyTime is a standardized hypermedia structuring language for representing hypertext linking, temporal and spatial event scheduling, and synchronization.

Although HyTime is an SGML application, it goes beyond SGML by specifying a way in which logically connected or time-related information can be described within the framework of SGML.

Links and Locations

HyTime provides two ways of linking:

The contextual link or *clink* is the simpler of the HyTime link constructs. One of its anchors ("the context") is part of the document where the link markup resides. The other endpoint can be in the same or some other document.

The independent link or *ilink* permits the link data to be stored externally, separate from the document(s) that the link markup connects. One can thus create links or make changes to the link data without modifying the documents being linked (Figure 2).

Both concepts bear important implications for their use in the domain of medical records. Since patient records are considered 'legally authenticated'

Figure 2: Independent links (in HyTime) or extended links (in XLink) are entities independent of and external to the linked resources

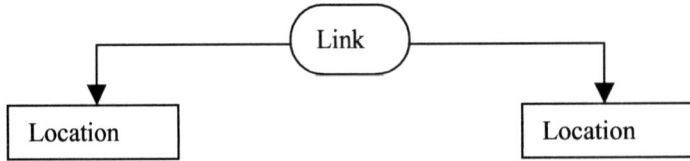

documents that carry an author and a creation date, and that are subject to the principle of integrity and longevity, some constraints on the linking technique have to be made. It is proposed that since links are part of the medical information, it follows that they belong integrally to the context of the 'resource documents.' With clinks this poses no problems, but if one would like to exploit the advantages of independent links, then awareness must be created that the documents containing the link data are also considered as part of the patient record with the same legal properties as their resource documents.

XLink

Like SGML, HyTime has so far not found many applications, probably because of its complexity. However, it serves as an ingenious concept that influences developments in the XML area. For XML to be useful in hypermedia applications there is a need for a linking specification. In December 2000 the W3C proposed a recommendation for an XML Linkage Language ('XLink'; DeRose, Maler, & Orchard, 2000). Like HyTime, XLink offers two kinds of links, which correspond to the concept of clink and ilink in HyTime.

Simple Links
Simple links offer shorthand syntax for a common kind of link, an outbound link with exactly two participating resources (into which category HTML-style A and IMG links fall). Because simple links offer less functionality than extended links, they have no special internal structure (Figure 3; DeRose et al., 2000).

Extended Links
Extended links offer full XLink functionality, such as inbound and third party arcs, as well as links that have arbitrary numbers of participating resources. As a result, their structure can be fairly complex, including elements for pointing to remote resources, elements for containing local resources, elements for specifying arc traversal rules, and elements for specifying human-readable resource and arc titles (Figure 4).

Figure 3: Simple links (in XLink) are outbound links with exactly two participating resources

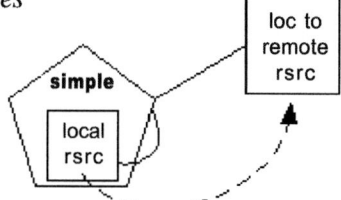

Figure 4: Extended links (in XLink) allow rich and complex ways to link arbitrary numbers of participating resources

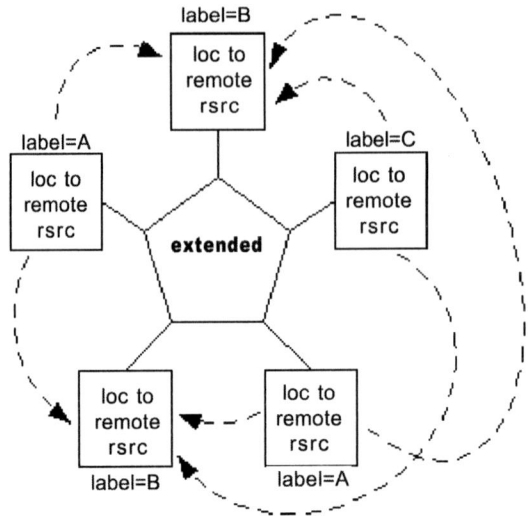

Extended links allow the inclusion of some semantic information:

XLink defines a way to give an extended link special semantics for finding linkbases; used in this fashion, an extended link helps an XLink application process other links.

The extended-type element may have the semantic attributes role *and* title. ... *They supply semantic information about the link as a whole; the* role *attribute indicates a property that the entire link has, and the* title *attribute indicates a human-readable description of the entire link* (DeRose et al., 2000).

Links are of course key ingredients for a Medical Data Web. But again, the usefulness of a linking concept largely depends on the ability to flexibly address structured data. The important idea is that fragments of a data structure can be located, for instance, by string matching, even if these locations have not been anticipated as linking anchors by the author of the target document.

For this purpose, an XML Pointer language (XPointer) has been specified by the W3C (DeRose, Maler, & Daniel, 2001). XPointer, which is based on the XML Path Language (XPath; Clark & DeRose, 1999), supports addressing the internal structures of XML documents. It allows for examination of a hierarchical document structure and choice of its internal parts based on various properties, such as element types, attribute values, character content, and relative position.

Beyond Hypertext–Topic Maps, Resource Description Framework and the Semantic Web

Powerful as the concepts and techniques of linking are, they carry a potential danger of overabundance of information. Traveling through a complex Web of data, one might well end lost–a phenomenon often experienced by roamers of the WWW. It is obvious that some kind of road map is needed to help navigate through the Medical Data Web. One possible solution to this problem is 'Topic Maps.'

Hypertext links belong to the context of the resources, even when they are handled in separate documents as ilinks or extended links. However, one can envisage a further level of abstraction by expressing connectivity at a meta-level above the resources. This concept forms the basis of a construct named Topic Maps (TMs). The TM standard has been developed on the basis of HyTime and has recently been given the status of an international standard (ISO/IEC 13250:1999; Biezunski, Bryan, & Newcomb, 1999).

A *topic* is created by linking a particular *topic name*, or a set of topic names, to one or more *occurrences* of references to the topic. References can be assigned different *occurrence roles* so that occurrences can be grouped according to their type. Both names and occurrences can be assigned *scopes* (domains) to distinguish them from similarly named/located topics.

Association links can be used to link topics into a navigable topic network. A set of topic links and topic associations defined in a separate resource forms a *topic map*.

Facet properties can be assigned to topics to allow different views of topics to be provided to different users, according to circumstances. So, for example, a French-speaking surgeon can have a different view of a set of information on a skin tumor than a German-speaking dermatologist does.

One of the advantages of TMs is their type structure. Because topics can be typed, it offers the possibility of introducing semantics. Not only topics can be typed, but so too can the relationships between topics. Such relationships are called associations, with association roles that can be typed. TMs allow creating strongly typed, linked models of an area of information. In this manner, the potential information glut created by the possible multitude of extended links and its unmanageably large number of possible pathways can be structured and made searchable. TMs allow overlying loosely structured data with a semantic model without changing the resource documents containing the data (Figure 5).

XML Topic Maps–XTM

TMs, as an application of the SGML-based HyTime standard, can already be expressed as XML documents. However, there is another endeavor, launched by an authoring group named 'TopicMaps.Org,' to define an XTM specification that

Figure 5: The Topic Map of a laboratory value

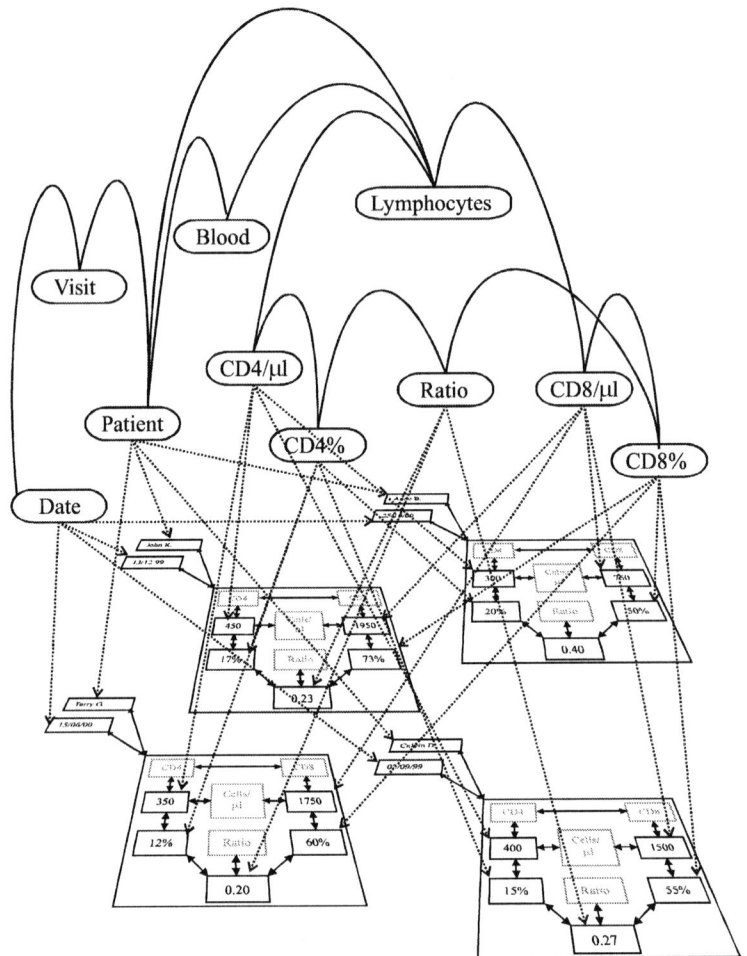

Navigation through a Web of clinical data is facilitated by a "map" of semantic information (associated topics) on a meta-level pointing to individual occurences of topics within a particular scope. The roles and types of occurences (...) and associations (--) are not indicated.

constrains the full generality of the TM standard in order to more easily exploit TMs on the WWW. The first version of XTM was made public in December 2000 by Biezunski et al.

Resource Description Framework (RDF)

The concept of TMs has many similarities to an XML-based specification called Resource Description Framework (RDF) recommended by the W3C in 1999 (Lassila & Swick, 1999) for describing data contained on the WWW. The basic data model of RDF is a statement in the form of the triplet *subject*, *predicate* and *object* corresponding to *resource*, *property* and *value*. Properties and values of the triplet, as well as the whole statement, can be resources of the Web and can have their own properties. RDF provides a framework in which independent communities can develop vocabularies that suit their specific needs and share vocabularies with other communities. The descriptions of these vocabulary sets are called RDF Schemas (Brickley & Guha, 2000).

The exact relationship between TMs and RDF is currently a matter of debate. One interesting possible outcome of these discussions is that the XTM syntax might be developed in a way that it can also be used for the interchange of RDF statements.

Other similarities exist between RDF and XLink because both provide a way of describing relationships between resources. XLink links can be mapped to RDF statements, that is, RDF statements can be harvested from XLinks (Daniel, 2000).

The Semantic Web Initiative of W3C

All these ingredients of modern Web technology that have been described above might develop into something called the *Semantic Web*. The Semantic Web lies at the heart of Tim Berners-Lee's vision for the future of the WWW: "The idea of having data on the Web defined and linked in a way that it can be used by machines not just for display purposes, but for automation, integration and reuse of data across various applications" (Miller, 2001; Introduction)

The important cornerstones in the work-in-progress that is to become the Semantic Web are the RDF data model and RDF schemas and their spin-offs. For example, semantic information can be attached to pictures by embedding RDF content in the comment portion of JPEG files (Lafon & Bos, 2000). This is one possible way to include pictorial data in the Medical Data Web.

Another example is the employment of RDF in the Composite Capabilities/Preference Profiles (CC/PP) to describe the capabilities and preferences of the Web clients. CC/PP is an extension of the HTTP protocol that will allow the client to inform the server about its properties (Reynolds, Hjelm, Dawkins, & Singhal, 1999). This feature will be very important in the Medical Data Web, given that the ease of access to medical information on different devices like mobile phones, handheld computers, or desktops is pivotal to the success of the Medical Data Web.

Of course, there will be multiple other ways to exploit the Semantic Web for medical information and, as such, the Medical Data Web will be just one specific domain of the universal Semantic Web.

DISCUSSION

It seems for the first time that the conceptual and technical solutions are available which will satisfy the requirements of a Medical Data Web. The ideas and solutions coming from the WWW are developing rapidly, driven by forces found in industry and the electronic business world as well as in academic spheres. The healthcare domain will be just one of many areas in which similar developments will take place. The question is not only how well and how quickly the healthcare domain will integrate the emerging techniques and how much it can profit from them, but also how much it can help to shape the future of the concepts.

Fulfillment of Requirements for a Medical Data Web

We have put forward two basic concepts, the hypertext and the markup paradigms, as solutions for the connectivity and structuring problem of medical information processing. How well do the technical realizations of these concepts promise to fulfill the requirements of a Medical Data Web?

1) Granularity can be well achieved by structuring medical data with XML markup. Because of its hierarchical structure, XML allows for an adaptable degree of granularity that can be dynamically adjusted to the needs of the application and the user.

2) The attachment of semantic information to data, links, and structures is well supported by the different specifications within the XML framework. Already HTML provides a *class* attribute (used most often with CSS stylesheets) that can contain semantic information. In XML, data can be typed with XML schemas; extended links of the XLink specification may have the semantic attributes *role* and *title*; TMs provide occurrence roles as well as topic types and association roles and types; and RDF's data model uses triplets of subject, predicate and object to generate semantic statements.

3) The constructs of XLink will allow a standard way of expressing and storing hypertext properties of medical information. On top of that, TMs will allow for a standard way of exchanging and storing medical concepts with well-defined semantics. Together, XLink and TMs (and/ or RDF) will be well-suited to handle the connectivity of medical data.

4) XML and TM-related query languages are still under development. The success of applying these standards for medical data management will largely depend on how well these query languages will support intuitive and efficient data retrieval. Some important requirements in relation to data structures and connectivity are specified by theW3C (Chamberlin & Fankhauser, et al., 2001):

> • *Queries MUST support operations on hierarchy and sequence of document structures.*
> • *Queries MUST be able to preserve the relative hierarchy and sequence of input document structures in query results.*
> • *Queries MUST be able to transform XML structures and MUST be able to create new structures.*
> • *Queries MUST be able to traverse intra- and inter-document references.*

5) XSL and SVG are important ingredients for an adaptable, dynamic, and user-friendly way of displaying medical data and their connections. However, to navigate efficiently through an extensive web of TM-based medical concepts and data, further tools will be necessary. Ideas for such instruments have been sketched out, but further research is certainly necessary in this area. A list of various Web visualization projects has been compiled at the CyberStacks(sm) site (McKiernan & Wasilko, 1999).

In summary, the instruments for weaving a Medical Data Web are out there. Nevertheless, although the concepts of hypertext and markup have been around for some time, the technical practicability of the concepts has only recently reached a level that allows their widespread use. How far has the healthcare sector followed this evolution?

Adoption of the Markup Paradigm in Healthcare

Some major healthcare-related organizations are definitively moving towards adopting XML as a language for storing and exchanging medical documents and data. One of them, the CDA framework of HL7, has been discussed above.

Following the United Kingdom government's decision to use XML for its Government Interoperability Framework (IAGC Data Interoperability Working Group, 2000), the National Health System in the United Kingdom is running several projects (Roberts et al., 2000) to use XML in their healthcare system. These projects include: XMedicaL, based on the Oswestry pilot project (Leeming et al., 1999) using SGML/XML for medical records, and the Scottish Immediate Discharge Document Project (SHOW-ISD Scotland, 2000) for automated XML discharge documents.

But not only patient records are moving towards XML—the framework of scientific literature and guidelines is moving this way too.

The U.S. National Library of Medicine (NLM, 2000) already employs XML for MEDLINE, which has approximately 8,000 references added each week. From 2001, XML will be the only distribution format for MEDLINE data. XML is also the format with which publishers will send bibliographic data. In addition, NLM has plans to produce an XML version of its controlled vocabulary Medical Subject Headings (MeSH), which is used to index the articles. NML is firmly committed to the use of XML. This is a very influential step since practically all the medical scientific literature is represented in MEDLINE.

Scientific medical literature is the basis not only for further scientific work but also for evidence-based and rational-guided medical decisions. Medical guidelines can also use XML, as proposed in a Guideline Elements Model (Shiffman, Agraval, Karras, Marenco, & Nath, 2000). A special tool *GEM Cutter* has been developed to facilitate the transformation of guideline information into GEM format. The standardized structuring of guidelines by using XML promises to facilitate the connection of that type of information with the actual patient data in a Medical Data Web.

Adoption of the Hypertext Paradigm in Healthcare

In comparison to the endeavors to introduce XML for structuring medical data, the hypertext paradigm has not yet been fully embraced in the medical domain, with the exception of scientific literature citation services. Here however, the hypertext concept has been embraced rapidly in recent years. PubMed, one of the two NLM providers of MEDLINE's 10,000,000 biomedical journal citations, has established an interesting and very useful system. Each citation contains a link to related articles about the same subject. In this way a huge hypertext Web of citations is created. LinkOut, another service of PubMed, provides links created by publishers, aggregators, libraries, biological databases, sequence centers and other Web resources. These links can take users to a provider's site to obtain the full text of articles, or related resources, e.g., protein structures or consumer health information. A publisher who wants to participate in LinkOut has to provide two XML files using the document type definition of LinkOut. The first file provides information on the publisher; the second describes the electronic journal holdings that are provided by a publisher. As of February 17, 2001, there are 213 resources participating in LinkOut with 1,702 journals providing links to full-text access to their articles.

Apart from these citation services, scientific articles themselves do not yet regularly contain hypertext links, nor do patient records containing medical data. HTML, the hypertext language of the WWW, is not suited for handling complex linking. However, with the advent of XML linking specifications, the door is open for developments of tools that are able to process linking information. Furthermore, the implementation of XTM and/or RDF specifications will eventually lead to information processing systems that can cope with semantics.

Integration of Medical Data Webs–
Transition from Old to New

When adopting new paradigms, an important concern is the transition from the old to the new. That part of information that resides in the structure and connectivity of medical data has been handled, intuitively, by the physician. Exchange and storage of such contextual information has been in the form of free text, be it in patient records or in scientific papers and guidelines. The Medical Data Web, as discussed above, will provide an explicit way of expressing such information. Of course, such significant change of paradigm will take its time to implement and will very likely never be complete. It is, therefore, important that the new concepts integrate smoothly and seamlessly into the classical, text-based systems. Both hypertext and markup technologies are well-suited for integration into a document-oriented environment. In fact, it is the background from where they originated. Critical, however, for the success of the Medical Data Web will be its query capabilities and the user interface: How can one get the right information at the right time and place in a useful way? Studying the way that a physician gathers information by the classical approach will likely help to develop a user interface that allows a seamless transition to the new facilities of a Medical Data Web.

Scenario Envisaged

What will the world of a Medical Data Web look like? Most likely, it will only gradually change the present way of delivering healthcare. Obviously, the work of a physician will still be based on well-established and proven foundations, but it will increasingly be enriched with and supported by more secure, quicker, and more specific access to information, both on the patient's status and risks as well as global medical knowledge. The medical decision process will be supported by automated alerts and reminders, by suggestions for diagnostic procedures and therapeutic measures, and by references to medical guidelines and scientific articles. The patient will be better informed about his disease, its prognosis and about the benefits and risk of intervention, and he will get timely and detailed guidance for adjusting his lifestyle. The monitoring of a chronic disease or long-term treatment will be improved and the compliance of the patient enhanced. Furthermore, statistical data on groups of patients, diseases and treatments will be easier to gather and to evaluate. This list of possible benefits is by no means exhaustive, and certainly there will also be risks involved that will have to be dealt with as the Medical Data Web evolves.

The gradual change in certain aspects of the doctor's daily work will be paralleled by a stepwise change in the technical machinery behind the scene. By way of illustration, let us try to envisage a scenario related to Figure 1. From a medical user's point of view, the example may appear trivial, but the underlying Medical Data Web is already fairly complex (Table 1).

CONCLUSION

Medical information processing by machines as an employment of knowledge media in healthcare depends on the accessibility of medical information in electronic form. The manner in which such information is expressed and perceived, both by computers and by human beings, is fundamental to the success of any attempt to profit from modern information technology in healthcare. It was argued that medical information resides not, to a large extent, in the data content of information itself but is rather distributed within the connections (links), as well as within the structured

Table 1: A 10-minute scenario

A 10-minute scenario	Behind the scene
Between 2 and 3 p.m., a time scheduled for non-planned but non-urgent access to the medical doctor, her mobile phone rings and a computer voice tells her that an anomalous change in the viral load values of an HIV patient has occurred. She takes her handheld computer and sees on the graphical display the unexpected increase of the viral load, compared to earlier checks.	The lab analyzer performs a quantitative HIV-RNA determination and the result is sent to the lab system using the SyncML protocol. There, it is stored in an XML structure and typed as a logarithmic value of RNA copies per milliliter by an XML schema. The result is linked to the patient by using XLink.
Since she cannot believe this change, she uses the same display to check the recent CD4 values of the patient and the treatment regiment, which show no alterations. However, the computer gives her a hint by suggesting that she look at the screen that displays the monitoring of the drug intake by the patient. These data are collected automatically when the patient opens his bottle of drugs, and they are sent wirelessly to the clinic. There, she discovers a recent lack of compliance of the patient, and she finds on the list of side effects, regularly reported and automatically transferred from the patient's mobile device, an increase of nausea that probably led to the reduced compliance.	The lab system using an XML query automatically compares the value with earlier results and detects a significant deviation. It triggers a voice message dispatch coded in VoxML that alerts the mobile phone of the doctor. This answers with an HTTP GET request containing a CCPP/RDF for voice. The voice message alerts the doctor, who activates her handheld computer, which sends its own CCPP for colour graphical display, and the relevant data are shown in an SVG graph. The viral load data are linked to the CD4 and CD8 values by XLink.
With a click on the screen, she arranges a visit for the patient to discuss with him a new treatment regiment. In the meantime, she goes to her desktop where	The lab data are translated by XSLT and sent to the decision support system of the clinic, which associates the data with the drug monitoring data of the patient using an XTM-based system. The drug intake data have been collected and sent by the

Table 1: A 10-minute scenario (continued)

where she finds, with a few mouse clicks, a guideline recommending some possible changes in treatment, given the actual clinical and laboratory data of the patient. Since these recommendations are new to her, she follows a link to a recent review of the scientific literature on that subject. At the same time, she provides some of the anonymized patient's data to a repository of new side effects for that particular drug. Soon thereafter, the repository returns an e-mail to her indicating that a special subgroup of patients is susceptible to this particular side effect. With another mouse click she creates a short summary of the event for the patient's record, edits it shortly and signs it electronically.	drug bottle device using SyncML syntax. The decision support system sends the collected information to the handheld computer of the doctor together with side effect data that have been collected by the patient using a WAP system on his mobile phone. To arrange a visit for the patient, the doctor triggers a request to the patient management system, which contacts the patient by e-mail. From her desktop the doctor uses an evidence-based medicine system that takes the actual patient data as an input and delivers a selection of guidelines in GEM syntax and links to further literature. The side effect data are delivered to the repository by using the CDISC DTD. A summary of the event is produced by an intelligent aggregation system using XTM/RDF, and the report is documented in the patient record using the CDA syntax and signed with an 'XML digital signature.'

context of the data elements. To enable a computerized processing of such information, some basic requirements for a structured and linked data model have to be fulfilled:

1. Data elements have to be granular enough to identify and address relevant atomic units of data.
2. There has to be a standardized way to attach semantic information to the data elements, links and structures.
3. The granular structure and the connections between the data elements have to be stored together with the data in a standardized way in a database.
4. A query system has to be able to parse the structure and links and to extract the information contained within the connectivity as well as from the data proper.

5. The query result has to be displayed for the user in a way that structure and connectivity are intuitively and usefully expressed and can be stored again in a structured, machine-accessible way.

This chapter argues that the developing Internet technology, as so successfully expressed in the WWW, provides a suitable model for how these requirements can be fulfilled. The hypertext paradigm together with the markup technology for structuring information provides all the necessary ingredients for developing information networks that might be called Medical Data Webs. On top of that, TMs and RDFs can be employed to semantically navigate through these Medical Data Webs. As such, Medical Data Webs will be just particular expressions of the general concept of the Semantic Web. The exact shape and functionality of Medical Data Webs are a matter of current and future developments are hard to predict. However, they will certainly be influenced by similar developments in other Web domains, such as e-business, and Medical Data Webs will have a reciprocal influence on the evolution in other domains of the Semantic Web.

ENDNOTE

1 Terms defined in the glossary are identified with single quotes at the first mention. Acronyms are expanded also in the glossary.

REFERENCES

Adler, S., Berglund, A., Caruso, J., Deach, S., Grosso, P., Gutentag, E., Milowski, A., Parnell, S., Richman, J. and Zilles, S. (2000). *Extensible Stylesheet Language (XSL) Specification Version 1.0. W3C Candidate Recommendation 21 November 2000*. Retrieved February 17, 2001 on the World Wide Web: http://www.w3.org/TR/WD-xsl.

Adobe®. (2000). *SVG Viewer Release 1.0*. Retrieved October 21, 2000 on the World Wide Web: http://World Wide Web.adobe.com/svg.

Alschuler, L. (1998). *ABCD... SGML: A User's Guide to Structured Information*. London/Boston: International Thomson Computer Press (ITCP).

Alschuler, L., Dolin, R. H., Boyer, S., Beebe, C., Biron, P. V. and Sokolowski, R. (2000). *Clinical Document Architecture Framework Version 1.0 Draft*. Retrieved October 21, 2000 on the World Wide Web: http://www.hl7.org/Library/Committees/structure/CDAMemberBallot.zip.

Bartel, M., Boyer, J., Fox, B. and Simon, E. (2000). *XML-Signature Syntax and Processing. W3C Candidate Recommendation 31-October-2000*.

Retrieved February 17, 2001 on the World Wide Web: http://www.w3.org/TR/xmldsig-core/.

Berners-Lee, T. (1989). *Information Management: A Proposal.* Retrieved October 21, 2000 on the World Wide Web: http://www.w3.org/History/1989/proposal.rtf.

Biezunski, M., Bryan, M. and Newcomb, S. (1999). *ISO/IEC FCD 13250:1999-Topic Maps.* Retrieved October 21, 2000 on the World Wide Web: http://www.ornl.gov/sgml/sc34/document/0058.htm and http://www.y12.doe.gov/sgml/sc34/document/0129.pdf.

Biezunski, M. and Newcomb, S. et al. (2000). *XTM: XML Topic Maps.* TopicMaps.Org AG Review Specification 4 Dec 2000. Retrieved February 17, 2001 from the World Wide Web: http://World Wide Web.topicmaps.org/xtm/1.0/.

Biron, P. V. and Malhotra, A. (2000). *XML Schema Part 2: Datatypes. W3C Candidate Recommendation 24 October 2000.* Retrieved October 24, 2000 on the World Wide Web: http://www.w3.org/TR/xmlschema-2.

Bray, T. (1998). *Introduction to the Annotated XML Specification: Extensible Markup Language (XML) 1.0.* Retrieved October 21, 2000 on the World Wide Web: http://www.xml.com/axml/testaxml.htm.

Bray, T., Paoli, J., Sperberg-McQueen, C. M. and Maler E. (2000). *Extensible Markup Language (XML) 1.0 (2nd Ed.). W3C Recommendation 6 October 2000.* Retrieved October 21, 2000 from the World Wide Web: http://www.w3.org/TR/REC-xml.

Brickley, D. and Guha, R. V. (2000). *Resource Description Framework (RDF) Schema Specification 1.0. W3C Candidate Recommendation 27 March 2000.* Retrieved October 21, 2000 on the World Wide Web: http://www.w3.org/TR/rdf-schema.

Bush, V. (1945). *As We May Think. The Atlantic Monthly.* HTML version by Duchier, D. (1994). Retrieved October 21, 2000 on the World Wide Web: http://www.csi.uottawa.ca/~dduchier/misc/vbush/awmt.html.

Chamberlin, D., Fankhauser, P., Marchiori, M. and Robie, J. (2001). *XML Query Requirements. W3C Working Draft 15 February 2001.* Retrieved February 17, 2001 on the World Wide Web: http://www.w3.org/TR/xmlquery-req.

Chamberlin, D., Florescu, D., Robie, J., Siméon, J. and Stefanescu, M. (2001). *XQuery: A Query Language for XML. W3C Working Draft 15 February 2001.* Retrieved February 16, 2001 on the World Wide Web: http://www.w3.org/TR/xquery.

Clark, J. (1999). *XSL Transformations (XSLT) Version 1.0. W3C Recommendation 16 November 1999.* Retrieved October 21, 2000 on the World Wide Web: http://www.w3.org/TR/xslt.

Clark, J. and DeRose, S. (1999). *XML Path Language (XPath). Version 1.0. W3C Recommendation 16 November 1999*. Retrieved October 21, 2000 on the World Wide Web: http://www.w3.org/TR/xpath.

Cover, R. (2001). *The XML Cover Pages: Proposed Applications and Industry Initiatives*. Retrieved February 10, 2001 on the World Wide Web: http://www.oasis-open.org/cover.

Daniel, R. Jr. (2000). *Harvesting RDF statements from XLinks. W3C Note 29 September 2000*. Retrieved October 10, 2000 on the World Wide Web: http://www.w3.org/TR/xlink2rdf.

DeRose, S., Maler, E. and Daniel, R. Jr. (2001). *XML Pointer Language (XPointer) Version 1.0. W3C Last Call Working Draft 8 January 2001*. Retrieved February 17, 2001 on the World Wide Web: http://www.w3.org/TR/xptr.

DeRose, S., Maler, E. and Orchard, D. (2000). *XML Linking Language (XLink) Version 1.0. W3C Proposed Recommendation 20 December 2000*. Retrieved February 17, 2001 on the World Wide Web: http://www.w3.org/TR/xlink.

Ferraiolo, J. et al. (2000). *Scalable Vector Graphics (SVG). 1.0 Specification. W3C Candidate Recommendation 02 November 2000*. Retrieved February 17, 2001 on the World Wide Web: http://www.w3.org/Graphics/SVG/Overview.htm8.

Fierz, W. and Grütter, R. (2000). The SGML standardization framework and the introduction of XML. *Journal of Medical Internet Research*, 2(2), e12. Retrieved October 21, 2000 on the World Wide Web: http://www.jmir.org/2000/2/e12/.

Goldfarb, C. F. (1990). *The SGML Handbook*. Oxford, England: Clarendon Press.

Goldfarb, C. F., Newcomb, S. R., Kimber, W. E. and Newcomb, P. J. (1997). *Information Processing—Hypermedia/Time-based Structuring Language (HyTime)-2d edition*. ISO/IEC 10744:1997. Retrieved October 21, 2000 on the World Wide Web: http://www.ornl.gov/sgml/wg8/docs/n1920.

Health Level Seven (2000). *Reference Information Model. Version 1.0*. Retrieved February 17, 2001 on the World Wide Web: http://www.hl7.org/library/data-model/RIM/C30100/rim0100h.htm.

IAGC Data Interoperability Working Group. (2000). *UK Government Interoperability Framework, Draft Version: 0.4 for global consultation*. Retrieved October 21, 2000 on the World Wide Web: http://www.citu.gov.uk/interoperability.doc.

Lafon, Y. and Bos, B. (2000). *Describing and retrieving photos using RDF and HTTP. W3C Note, 28 September 2000.* Retrieved February 15, 2001 on the World Wide Web: http://www.w3.org/TR/photo-rdf/.

Lassila, O. and Swick, R. R. (1999). *Resource Description Framework (RDF) Model and Syntax Specification. W3C Recommendation 22 February 1999.* Retrieved October 21, 2000 on the World Wide Web: http:// www.w3.org/TR/REC-rdf-syntax

Leeming, J., Garnett, D. and Roberts, A. (1999). *The 'Oswestry' EPR Project (1994-1999) A Final Report.* Retrieved October 21, 2000 on the World Wide Web: http://www.nhsia.nhs.uk/erdip/documents/OswestryFinal.doc.

McKiernan, G. and Wasilko, P. J. (1999). *Big Picture(sm): Visual Browsing in Web and non-Web Databases.* Retrieved October 21, 2000 on the World Wide Web: http://www.iastate.edu/~CYBERSTACKS/ BigPic.htm.

Miller, E. (2001). *W3C Semantic Web Activity Statement.* Retrieved February 15, 2001 on the World Wide Web: http://www.w3.org/ 2001/sw/Activity.

NLM, U.S. National Library of Medicine. (2000). *Information for Licensees of NLM Data.* Retrieved October 21, 2000 on the World Wide Web: http://www.nlm.nih.gov/news/medlinedata.html.

Reynolds, F., Hjelm, J., Dawkins, S. and Singhal, S. (1999). *Composite Capability/Preference Profiles (CCPP): A User Side Framework for Content Negotiation. W3C Note 27 July 1999.* Retrieved February 15, 2001 on the World Wide Web: http://www.w3.org/TR/NOTE-CCPP/.

Roberts, A. et al. (2000). *XMedicaL.* Retrieved October 21, 2000 on the World Wide Web: http://www.xmedical.org.uk/.

Robie, J., Chamberlin, D. and Florescu, D. (2000). Quilt: An XML query language. In *Proceedings of XML Europe. Graphic Communications Association.* Retrieved October 21, 2000 on the World Wide Web: http:// www.gca.org/papers/xmleurope2000/papers/s08-01.html.

Shiffman, R. N., Agraval, A., Karras, B., Marenco, L. and Nath, S. (2000). *GEM: The Guideline Elements Model. Yale Center for Medical Informatics.* Retrieved October 21, 2000 on the World Wide Web: http:// ycmi.med.yale.edu/GEM/.

SHOW-ISD Scotland. (2000). *Scottish Immediate Discharge Document Project.* Retrieved October 21, 2000 on the World Wide Web: http://scotland-xml.uk.eu.org/iddproject.html.

Thompson, H. S., Beech, D., Maloney, M. and Mendelsohn, N. (2000). *XML Schema Part 1: Structures. W3C Candidate Recommendation 24 October 2000*. Retrieved October 24, 2000 on the World Wide Web: http://www.w3.org/TR/xmlschema-1/.

World Wide Web Consortium (W3C). (2000). *Members*. Retrieved October 21, 2000 on the World Wide Web: http://www.w3.org/Consortium/Member/List.html.

GLOSSARY

Term or Abbreviation	Definition
Architecture	An architecture for structured documents defines relationships between documents and document specifications in terms of specialization and inheritance (see also CDA architecture)
CDA	Clinical Document Architecture (Alschuler, et al., 2000)
Clinical Document Architecture (CDA)	An XML application that specifies the structure and semantics of clinical documents for the purpose of exchange
CDISC[2]	Clinical Data Interchange Standards Consortium
Data	Atomic units of information
Document type definition (DTD)	A grammar for a class of documents; a type of SGML/XML schema
DTD[2]	Document Type Definition
GEM[2]	Guideline Elements Model
Granularity	The relative size of a defined addressable unit
HL7	Health Level 7, Inc. (http://www.hl7.org)
HTML[2]	Hypertext Markup Language, an SGML-based specification of the W3C
HyTime[2]	Hypermedia/Time-based Structuring Language, ISO/IEC 10744:1997 (Goldfarb et. al., 1997)
Hypertext Hypermedia	Presentation of information as a linked network of nodes that can be navigated freely
Information	Collection of data that are put into context
Legally authenticated	A completion status in which a document has been signed manually or electronically by the individual who is legally responsible for that document
Markup	Computer-processable annotations within a multimedia document
Medical Data Web	A structured and connected collection of data elements expressing medical information
MEDLINE	Medical Literature, Analysis, and Retrieval System Online
Quilt	An XML Query Language for heterogeneous data sources (Robie et al., 2000)
Resource	A document or database containing the locations that are to be linked
RIM	Reference Information Model of HL7 (Health Level Seven, 2000)
Schema	A formal definition of the structure and content of a class of documents
Semantic	The meaning of an element as distinct from its syntax
SGML[2]	Standard Generalized Markup Language, ISO 8879:1986 (Goldfarb, 1990)
SyncML[2]	XML for synchronizing all devices and applications over any network[2]
Topic Map (TM)	ISO/IEC 13250:1999 (Biezunski et al., 1999)
TopicMaps.Org	Authoring group for development of XML Topic Maps (XTM) (http://www.topicmaps.org)
VoxML[2]	Voice Extensible Markup Language
WAP[2]	Wireless Access Protocol (http://www.wapforum.org/)
Web, World Wide Web (WWW)	Global document and data network based on Internet technology and the Hypertext Transfer Protocol (HTTP)
W3C[2]	World Wide Web Consortium (2000), an international industry consortium
XLink[2]	XML Linkage Language (DeRose et al., 2000)
XML[2]	Extensible Markup Language, a formal subset of SGML (Bray, 1998)
XML digital signature[2]	Specification for integrity, message authentication, and/or signer authentication services for data of any type (Bartel, Boyer, Fox, & Simon, 2000)
XPath[2]	XML Path Language, a language for addressing parts of an XML document, designed to be used by both XSLT and XPointer (Clark & DeRose, 1999)
XPointer[2]	XML Pointer Language based on XPath supports addressing into the internal structures of XML documents (DeRose et al., 2001)
XSL[2]	Extensible Stylesheet Language (Adler et al., 2000)
XSLT[2]	XSL Transformations, a language for transforming XML documents into other XML documents (Clark, 1999)

[2] Further links to these resources can be found at the XML Cover Pages (Cover, 2001).

Chapter XIII

An Extensible Approach for Modeling Ontologies in RDF(S)

Steffen Staab, Michael Erdmann, Alexander Maedche
and Stefan Decker
University of Karlsruhe, Germany

INTRODUCTION

The development of the World Wide Web is about to mature from a technical platform that allows for the transportation of information from sources to humans (albeit in many syntactic formats) to the communication of knowledge from Web sources to machines. The knowledge food chain has started with technical protocols and preliminary formats for information presentation (HTML–HyperText Markup Language) over a general methodology for separating information contents from layout (XML–eXtensible Markup Language, XSL–eXtensible Stylesheet Language) to reach the realms of knowledge provisioning by the means of RDF and RDFS.

RDF (Resource Description Framework) is a W3C recommendation (Lassila & Swick, 1999) that provides description facilities for knowledge pieces, viz., for triples that denote relations between pairs of objects. To exchange and process RDF models they can be serialized in XML. RDF exploits the means of XML to allow for disjoint namespaces, linking and referring between namespaces and, hence, is a general methodology for sharing machine-processable knowledge in a distributed setting. On top of RDF the simple schema language *RDFS* (Resource Description Framework

Schema; Brickley & Guha, 1999) has been defined to offer a distinguished vocabulary to model class and property hierarchies and other basic schema primitives that can be referred to from RDF models. To phrase the role of RDFS in knowledge engineering terminology, it defines a simple *ontology* that particular RDF documents may be checked against to determine consistency.

Ontologies have shown their usefulness in application areas such as intelligent information integration or information brokering. Therefore their use is highly interesting for Web applications, which may also profit from long-term experiences made in the knowledge acquisition community. At the same time, this is a great chance for the knowledge acquisition community as RDF(S) may turn knowledge engineering, so far a niche technology, into a technological and methodological powerhouse. Nevertheless, while support for modeling of ontological concepts and relations has been extensively provided in RDF(S), the same cannot be said about the modeling of ontological axioms—one of the key ingredients in ontology definitions and one of the major benefits of ontology applications.

RDF(S) offers only the most basic modeling primitives for ontology modeling. Even though there are good and bad choices for particular formal languages, one must face the principal trade-off between tractability and expressiveness of a language. RDF(S) has been placed nearer to the low end of expressiveness, because it has been conceived to be applicable to vast Web resources! In contrast to common knowledge representation languages, RDF(S) has not been meant to be the definitive answer to all knowledge representation problems, but rather an *extensible core language*. The namespace and reification mechanisms of RDF(S) allow (communities of) users to define their very own standards in RDF(S) format—extending the core definitions and semantics. As RDF(S) leaves the well-trodden paths of knowledge engineering at this point, we must reconsider crucial issues concerning ontology modeling and ontology applications. To name but a few, we mention the problem of merging and mapping between namespaces, scalability issues, or the definition and usage of ontological axioms.

In this paper we concentrate on the latter, namely on how to model axioms in RDF(S) following the stipulations, *(i)* that the core semantics of RDF(S) is reused such that "pure" RDF(S) applications may still process the core object-model definitions, *(ii)* that the semantics is preserved between different inferencing tools (at least to a large stretch), and *(iii)* that axiom modeling is adaptable to reflect diverging needs of different communities. Current proposals neglect or even conflict with one or several of these requirements. For instance, the first requirement is violated by the ontology exchange language

XOL (Karp, Chaudhri, & Thomere, 1999) making *all* the object-model definitions indigestible for most RDF(S) applications. The interchangeability and adaptability stipulation is extremely difficult to meet by the parse-tree-based representation of MetaLog (Marchiori & Saarela, 1998), since it obliges to first-order logic formulae. We will show how to adapt a general methodology that we have proposed for axiom modeling (Maedche, Schnurr, Staab, & Studer, 2000; Staab & Maedche, 2000) to be applied to the engineering of ontologies with RDF(S). Our approach is based on translations of RDF(S) axiom specifications into various target systems that provide the inferencing services. As our running example, we map axiom specifications into an F-Logic format that has already served as the core system for SiLRi, an inference service for core RDF (Decker, Brickley, Saarela, & Angele, 1998). Our methodology is centered around categorization of axioms, because this allows for a more concise description of the *semantic meaning* rather than a particular syntactic representation of axioms. Thus, we get a better grip on extensions and adaptations to particular target inferencing systems.

In the following, we introduce the RDF(S) data model and describe how to define an object model in RDF(S) including practical issues of ontology documentation. Then we describe our methodology for using RDF(S) such that axioms may be engineered and exchanged. We describe the core idea of our approach and illustrate with several examples how to realize our approach. Before we conclude, we give a brief survey of related work.

MODELING CONCEPTS AND RELATIONS IN RDF(S)

In this section we will first take a look at the core ontology engineering task, i.e., at the RDF(S) data model proper, and then exploit RDF(S) also for purposes of practical ontology engineering, viz., for documentation of newly defined or reused ontologies. This will lay the groundwork for the modeling of axioms thereafter.

The RDF(S) Data Model

RDF(S) is an abstract data model that defines relationships between entities (called resources in RDF) in a similar fashion as semantic nets. Statements in RDF describe resources, which can be Web pages or surrogates for real-world objects like publications, pieces of art, persons, or institutions. We illustrate how concepts and relations can be modeled in RDF(S) by presenting a sample ontology in the abstract data model and only afterwards show how these concepts and relations are presented in the XML-serialization of RDF(S).

RDF

As already mentioned RDF(S) consists of two closely related parts: RDF and RDF Schema. The foundation of RDF(S) is laid out by RDF, which defines basic entities, like resources, properties, and statements. Anything in RDF(S) is a resource. Resources may be related to each other or to literal (i.e., atomic) values via properties. Such a relationship represents a statement that itself may be considered a resource, i.e., reification is directly built into the RDF data model. Thus, it is possible to make statements about statements. These basic notions can be easily depicted in a graphical notation that resembles semantic nets. To illustrate the possibilities of pure RDF the following statements are expressed in RDF and depicted in Figure 1, where resources are represented by rectangles, literal values by ovals and properties by directed, labeled arcs:

- Firstly, in part (a) of Figure 1 two resources are defined, each carrying a FIRSTNAME and a LASTNAME property with literal values, identifying the resources as William and Susan Smith, respectively. These two resources come with a URI as their unique global identifier and they are related via the property MARRIEDWITH, which expresses that William is married with Susan.

- Part (b) of the illustration shows a convenient shortcut for expressing more complex statements, i.e., reifying a statement and defining a property for the new resource. The example denotes that the marriage between William and Susan has been confirmed by the resource representing the Holy Father in Rome.

- The RDF data model offers the predefined resource `rdf:Statement` and the predefined properties `rdf:subject`, `rdf:predicate`, and `rdf:object` to reify a statement as a resource. The actual model for the example (b) is depicted in part (c) of Figure 1. Note that the reified statement makes no claims about the truth value of what is reified, i.e., if one wants to express that William and Susan are married *and* that this marriage has been confirmed by the pope then the actual data model must contain a union of part (a) and part (c) of the example illustration.

RDFS

As a companion standard to RDF, the schema language RDFS is more important with respect to ontological modeling of domains. RDFS offers a distinguished vocabulary defined on top of RDF to allow the modeling of

Figure 1: An example RDF data model

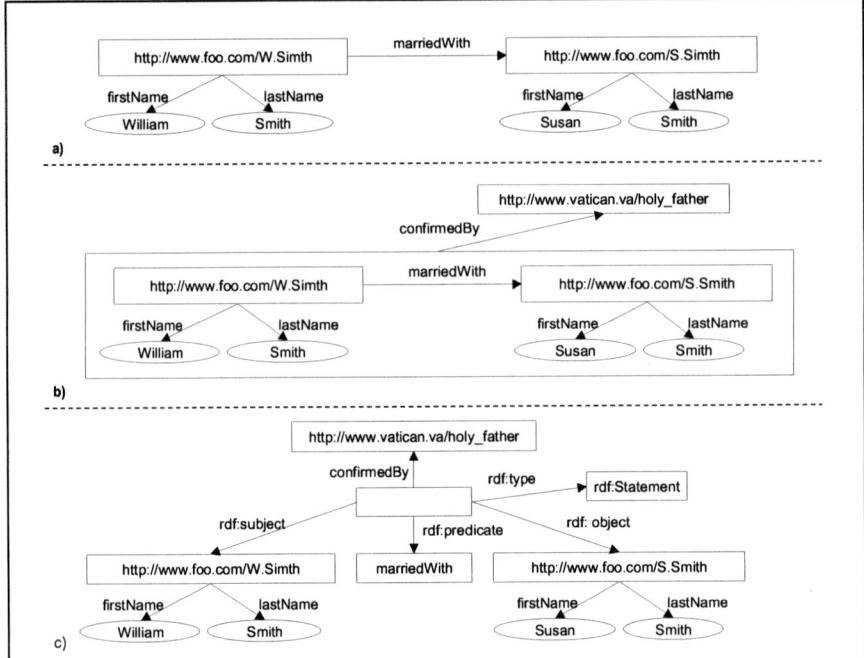

object models with cleanly defined semantics. The terms introduced in RDFS build the groundwork for the extensions of RDF(S) that are proposed in this chapter. The relevant RDFS terms are presented in the following list.

- The most general class in RDF(S) is `rdfs:Resource`. It has two subclasses, namely `rdfs:Class` and `rdf:Property` (cf. Figure 2). The reader may note that only a very small part of RDF(S) is depicted in the RDF/RDFS layer of the figure. Furthermore, the relation `appl:marriedWith` in the data layer is identical to the resource `appl:marriedWith` in the schema layer. When specifying a domain specific schema for RDF(S), the classes and properties defined in this schema will become instances of these two resources.

- The resource `rdfs:Class` denotes the set of all classes in an object-oriented sense. That means, that classes like `appl:Person` or `appl:Organisation` are instances of the meta-class `rdfs:Class`.

- The same holds for properties, i.e., each property defined in an application-specific RDF schema is an instance of `rdf:Property`, e.g., `appl:marriedWith`.

- RDFS defines the special property `rdfs:subClassOf` that defines the subclass relationship between classes. Since `rdfs:subClassOf` is transitive, definitions are inherited by the more specific classes from the more general classes, and resources that are instances of a class are automatically instances of all superclasses of this class. In RDF(S) it is prohibited that any class is an `rdfs:subClassOf` itself or of one of its subclasses.

- Similar to `rdfs:subClassOf`, which defines a hierarchy of classes, another special type of relation, `rdfs:subPropertyOf`, defines a hierarchy of properties, e.g., one may express that FATHEROF is a `rdfs:subPropertyOf` PARENTOF.

- RDFS allows one to define the domain and range restrictions associated with properties. For instance, these restrictions allow the definition that persons and only persons may be MARRIEDWITH and only with other persons.

As depicted in the middle layer of Figure 2 the domain specific classes `appl:Person`, `appl:Man`, and `appl:Woman` are defined as instances of `rdfs:Class`. In the same way domain-specific property types are defined as instances of `rdf:Property`, i.e., `appl:marriedWith`, `appl:firstName`, and `appl:lastName`.

The Use of XML Namespaces in RDF(S)

The XML namespace mechanism plays a crucial role for the development of RDF schemata and applications. It allows one to distinguish between different modeling layers (cf. Figures 2 and 3) and to reuse and integrate existing schemata and applications. At the time being, there exists a number of *canonical* namespaces, e.g., for RDF, RDFS, and Dublin Core. We here introduce two new namespaces that aim at two different objectives, viz., the comprehensive documentation of ontologies and the capturing of our proposal for the modeling of ontological axioms.

An actual ontology definition occurs at a concrete URL. The reader may actually compare with the documents that appear at these URLs, e.g. `http://ontoserver.aifb.uni-karlsruhe.de/schema/example.rdf`. It defines shorthand notations which refer to our actual namespaces for ontology documentation and modeling of ontological axioms, abbreviated `odoc` and `o`, respectively. An actual application that uses our example ontology will define a shorthand identifier like `appl` in order to refer to this particular, application-specific ontology. Figures 2 and 3 presume these shorthand notations for the namespaces we have just mentioned.

XML Serialization of RDF(S)

One important aspect for the success of RDF in the WWW is the way RDF models are represented and exchanged, namely via XML. In the

Figure 2: An example RDF schema and its embedding in RDF(S)

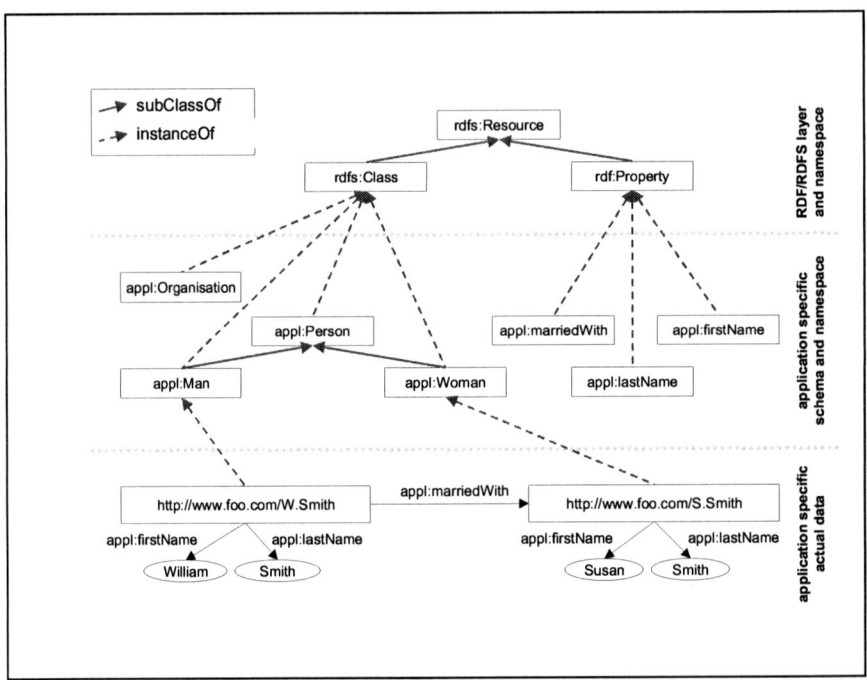

following excerpt of the RDF schema document http://ontoserver.aifb.uni-karlsruhe.de/schema/example.rdf, the classes and property types defined in Figure 2 are represented in XML and the domains and ranges of the properties are defined using the RDF constraint properties rdfs:domain and rdfs:range.

```
<rdf:Description ID="Person">
 <rdf:type resource="http://www.w3.org/TR/1999/PR-rdf-schema-
    19990303#Class"/>
</rdf:Description>

<rdf:Description ID="Man">
 <rdf:type resource="http://www.w3.org/TR/1999/PR-rdf-schema-
    19990303#Class"/>
 <rdfs:subClassOf rdf:resource="#Person"/>
</rdf:Description>

<rdf:Description ID="Woman">
 <rdf:type resource="http://www.w3.org/TR/1999/PR-rdf-schema-
    19990303#Class"/>
 <rdfs:subClassOf rdf:resource="#Person"/>
</rdf:Description>
```

```
<rdf:Description ID="Organisation">
 <rdf:type resource="http://www.w3.org/TR/1999/PR-rdf-schema-
    19990303#Class"/>
</rdf:Description>

<rdf:Description ID="firstName">
 <rdf:type resource="http://www.w3.org/TR/1999/PR-rdf-schema-
    19990303#Property"/>
 <rdfs:domain rdf:resource="#Person"/>
 <rdfs:range rdf:resource="http://www.w3.org/TR/xmlschema-2/
    #string"/>
</rdf:Description>

<rdf:Description ID="lastName">
 <rdf:type resource="http://www.w3.org/TR/1999/PR-rdf-schema-
    19990303#Property"/>
 <rdfs:domain rdf:resource="#Person"/>
 <rdfs:range rdf:resource="http://www.w3.org/TR/xmlschema-2/
    #string"/>
</rdf:Description>

<rdf:Description rdf:ID="marriedWith">
 <rdf:type resource="http://www.w3.org/TR/1999/PR-rdf-schema-
    19990303#Property"/>
 <rdfs:domain rdf:resource="#Person"/>
 <rdfs:range rdf:resource="#Person"/>
</rdf:Description>
```

Modeling Ontology Metadata Using RDF Dublin Core

Metadata about ontologies, such as the title, authors, version, statistical data, etc. are important for practical tasks of ontology engineering and exchange. In our approach we have adopted the well-established and standardized RDF Dublin Core Metadata element set (Weibel & Miller, 1998). This element set comprises 15 elements which together capture basic aspects related to the description of resources. Ensuring a maximal level of generality and exchangeability, our ontologies are labeled using this basic element set. Since ontologies represent a very particular class of resource, the general Dublin Core metadata description does not offer sufficient support for ontology engineering and exchange. Hence, we describe further semantic types in the schema located at http://ontoserver.aifb.uni-karlsruhe.de/schema/ontodoc and instantiate these types when we build a new ontology. The example below illustrates our usage and extension of Dublin Core by an excerpt of an exemplary ontology metadata description.

```
<?xml version='1.0' encoding='ISO-8859-1'?>
<rdf:RDF
 xmlns:rdf="http://www.w3.org/1999/02/22-rdf-syntax-ns#"
 xmlns:dc="http://purl.org/dc/documents/rec-dces-19990702.htm"
 xmlns:odoc="http://ontoserver.aifb.uni-karlsruhe.de/schema/
 ontodoc">
  <rdf:Description about="">
   <dc:Title>An Example Ontology</dc:Title>
   <dc:creator>
    <rdf:Bag>
     <rdf:li>Steffen Staab</rdf:li>
     <rdf:li>Michael Erdmann</rdf:li>
     <rdf:li>Alexander Maedche</rdf:li>
    </rdf:Bag>
   </dc:creator>
   <dc:date>2000-02-29</dc:date>
   <dc:format>text/xml</dc:format>
   <dc:description>An example ontology modeled for this small
    application</dc:description>
   <dc:subject>Ontology, RDF</dc:subject>
   <odoc:url>http://ontoserver.aifb.uni-karlsruhe.de/schema/
    example.rdf</odoc:url>
   <odoc:version>2.1</odoc:version>
   <odoc:last_modification>2000-03-01</odoc:last_
    modification>
   <odoc:ka_technique>semi-automatic text knowledge acqui-
    sition</odoc:ka_technique>
   <odoc:ontology_type>domain ontology</odoc:ontology_type>
   <odoc:no_concepts>24</odoc:no_concepts>
   <odoc:no_relations>23</odoc:no_relations>
   <odoc:no_axioms>11</odoc:no_axioms>
   <odoc:highest_depth_level>6</odoc:highest_depth_level>
  </rdf:Description>
</rdf:RDF>
```

MODELING OF AXIOMS IN RDF(S)

Having prepared the object-model and documentation backbone for ontologies in RDF(S), we may now approach the third pillar of our approach, viz., the specification of axioms in RDF(S). The basic idea that we pursue is the specification and serialization of axioms in RDF(S) such that they remain easily representable and exchangeable between different ontology engineering, representation and inferencing environments. The principal specification needs to be rather independent of particular target systems (to whatever extent this is possible at all) in order to be of value in a distributed Web setting with many different basic applications.

Axioms Are Objects, Too

Representation of interesting axioms that are deemed to be applied in different inferencing applications turns out to be difficult. The reason is that typically some kind of non-propositional logic is involved that deals with quantifiers and quantifier scope. Axioms are difficult to grasp, since the representation of quantifier scope and its likes is usually what the nitty-gritty details of a particular syntax, on which a particular inferencing application is based, are about. An ontology representation in RDF(S) should, however, abstract from particular target systems.

A closer look at the bread and butter issues of ontology modeling reveals that many axioms that need to be formulated aim at much simpler purposes than arbitrary logic structures. Indeed, we have found that many axioms in our applications belong to one of a list of major axiom categories:

1. Axioms for a relational algebra
   (a) Reflexivity of relations
   (b) Irreflexivity of relations
   (c) Symmetry of relations
   (d) Asymmetry of relations
   (e) Antisymmetry of relations
   (f) Transitivity of relations
   (g) Inverse relations
2. Composition of relations
3. (Exhaustive) Partitions
4. Axioms for subrelation relationships
5. Axioms for part-whole reasoning

Our principal idea for representing ontologies with axioms in RDF(S) is based on this categorization. The categories allow one to distinguish between the structures that are repeatedly found in axiom specifications from a corresponding description in a particular language. Hence, one may describe axioms as complex objects (one could term them instantiations of axiom schemata) in RDF(S) that refer to concepts and relations, which are also denoted in RDF(S). For sets of axiom types we presume the definition of different RDF schemata. Similar to the case of simple metadata structures, the RDF schema responsible for an axiom categorization obliges to a particular semantics of its axiom types, which may be realized in a number of different inferencing systems like description logics systems (e.g., Horrocks, 1998) or frame logic systems (Decker et al., 1998). The schema defined in our namespace http://ontoserver.aifb.uni-karlsruhe.de/schema/rdf stands for the semantics defined in this and our previous papers (Maedche et al., 2000; Staab & Maedche, 2000). The schema is also listed in the appendix of this paper. Other communities may, of course, find other reasoning schemes more important, or they may just need an extension compared to what we provide here.

Thus, we build a two-layer approach. On the first layer, the *symbol level*, we provide a RDF(S) syntax (i.e., serialization) to denote particular types of axioms. The categorization really constitutes a *knowledge level* that is independent from particular machines. In order to use an ontology denoted with our RDF(S) approach, one determines the appropriate axiom category and its actual instantiation found in a RDF(S) piece of ontology, translates it into a corresponding logical representation and executes it by an inferencing engine that is able to reason with (some of) the relevant axiom types.

Figure 3 summarizes our approach for modeling axiom specifications in RDF(S). It depicts the core of the RDF(S) definitions and our extension for axiom categorizations (i.e., our ontology meta-layer). A simple ontology, especially a set of application specific relationships, is defined in terms of our extension to RDF(S).

In the following subsections, we will further elucidate our approach by proceeding through a few simple examples of our categorization of axiom specifications listed above. In particular our scheme is *(A)* to show the representations of axioms in RDF(S) and *(B)* to show a structurally equivalent F(rame)-Logic representation that may easily be derived from its RDF(S) counterpart (cf. Kifer, Lausen, & Wu, 1995; Decker, 1998 on F-Logic). Then, *(C)* we exploit the expressiveness of F-Logic in order to specify translation axioms that work directly on the F-Logic object representation of axioms. Thus, *(B)* in combination with *(C)* describes a formally concise and executable translation. For better illustration, we finally, *(D)*, indicate the result of our translation by exemplary target representations of the axioms stated in RDF(S).

The reader should note here that we neither believe that F-Logic fulfills all the requirements that one might wish from an ontology inferencing language, nor do we believe that the axiom types we mention exhaust all relevant types. Rather we believe that our experiences in particular domains will push for further categorizations of axioms, further translation mechanisms, and, hence, further extensions of the core RDF(S) representation. All that will have to be agreed upon by communities that want to engineer and exchange ontologies with interesting axioms across particularities of inference engines. Our main objective is to acquaint the reader with our *principle methodology* that is transportable to other translation approaches, inferencing systems, and other axiom types, when need arises.

Figure 3: An object model and an instantiation in RDF(S)

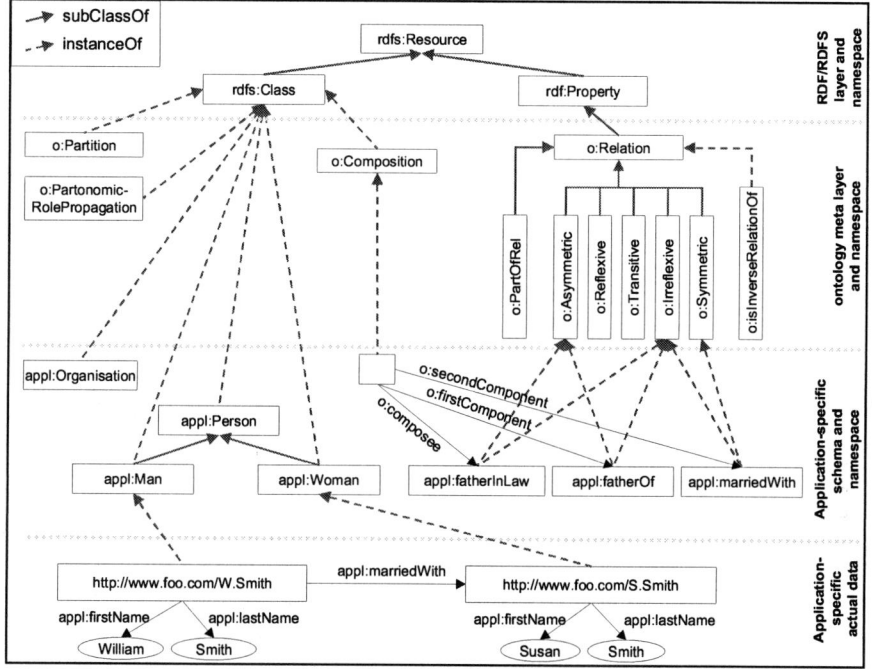

Axioms for a Relational Algebra

The axiom types that we have shown above are listed such that easier axioms come first and harder ones appear further down in the list. Axiom specifications that are referred to as "axioms for a relational algebra" rank among the simplest ones. They describe axioms with rather local effects, because their implications only affect one or two relations. We show here one simple example of these in order to explain the basic approach and some syntax. The principle approach easily transfers to all axiom types from 1.(a)-(g) to 5, as shown in the previous subsection.

Let us consider an example for symmetry. A common denotation for the symmetry of a relation MARRIEDWITH (such as used for "William is married with Susan") in first-order predicate logic boils down to:

(1) $\forall$ X, Y MARRIEDWITH$(X, Y) \leftarrow$ MARRIEDWITH(Y, X).

In F-Logic, this would be a valid axiom specification, too. Most often, however, modelers that use F-Logic take advantage of the object-oriented syntax. Concept definitions in F-Logic for *Person* having an attribute MARRIEDWITH and *Man* being a subconcept of *Person* are given in (2), while a fact that William is a *Man* who is MARRIEDWITH Susan appears like in (3).

(2) *Person*[MARRIEDWITH =>> *Person*].
   Man::Person.

(3)   William:*Man*[MARRIEDWITH –>> Susan].

Hence, a rule corresponding to (1) is given by (4).

(4) $\forall$ X, Y Y[MARRIEDWITH –>> X] $\leftarrow$ X[MARRIEDWITH –>> Y].

We denote symmetry as a predicate that holds for particular relations:

(5)   SYMMETRIC(MARRIEDWITH).

In RDF(S), this specification may easily be realized by a newly agreed upon class `o:Symmetric`:

(6)   `<o:Symmetric rdf:ID="marriedWith"/>`

For a particular language like F-Logic, one may then derive the implications of symmetry by a general rule and, thus, ground the meaning of the predicate SYMMETRIC in a particular target system. The corresponding transformation rule (here in F-Logic) states that if for all symmetric relations R and object instances X and Y it holds that X is related to Y via R, then Y is also related to X via R.

(7) $\forall$ R, X, Y Y[R –>> X] $\leftarrow$ SYMMETRIC(R) and X[R –>> Y].

This small example already shows three advantages. First, the axiom specification (6) is rather target-system independent. Second, it is easily realizable in RDF(S). Third, our approach for denoting symmetry is much sparser than its initial counterpart (4), because (7) is implicitly assumed as the agreed semantics for our schema definition.

Following our strategy sketched in the previous subsection, these steps from RDF representation to axiom meaning are now summarized in Table 1. For easier understanding, we also will reuse this table layout in the following subsection.

Composition of Relations

The next example concerns composition of relations. For instance, if a first person is FATHEROF a second person who is MARRIEDWITH a third person then one may assert that the first person is the FATHERINLAWOF the third person. Again different inferencing systems may require completely different realizations of such an implication. The object description of such an axiom may easily be denoted in F-Logic or in RDF(S) (cf. Table 2). The transformation rule works very similarly to the transformation rule for symmetry.

General Axioms

Our approach of axiom categorization is not suited to cover every single axiom specification one may think of. Hence, we still must allow for axioms that are specified in a particular language like first-order predicate logic and we must allow

Table 1: Symmetry

A	`<o:Symmetric rdf:ID="marriedWith"/>`	RDF(S)
B	SYMMETRIC (MARRIEDWITH)	F-Logic Predicate
C	$\forall\, R, X, Y\;\; Y[R \text{->>} X] \leftarrow$ SYMMETRIC (R) and $X[R \text{->>} Y]$.	Translation Axiom
D	$\forall\, X, Y\;\; X[\text{MARRIEDWITH} \text{->>} Y] \leftarrow Y[\text{MARRIEDWITH} \text{->>} X]$.	Target Axiom

Table 2: Composition

A	`<o:Composition rdf:ID="FatherInLawComp">`
	  `<o:composee rdf:resource="fatherInLawOf"/>`
	  `<o:firstComponent rdf:resource="fatherOf"/>`
	  `<o:secondComponent rdf:resource="marriedWith"/>`
	`</o:Composition>`

B	COMPOSITION(FATHERINLAWOF, FATHEROF, MARRIEDWITH)
C	$\forall\, R, Q, S, X, Y, Z\;\; X[S \text{->>} Z] \leftarrow$ COMPOSITION$(S, R, Q) \wedge X[R \text{->>} Y]$ and $Y[Q \text{->>} Z]$.
D	$\forall\, X, Y, Z\;\; X[\text{FATHERINLAWOF} \text{->>} Z] \leftarrow X[\text{FATHEROF} \text{->>} Y]$ and $Y[\text{MARRIEDWITH} \text{->>} Z]$.

for their representation in RDF(S). There are principally two ways to approach this problem. First, one may conceive a new RDF(S) representation format that is dedicated to a particular inferencing system for reading and performing inferences. This is the way that has been choosen for OIL (Horrocks et al., 2000), which has a RDF(S) style representation for a very simple description logics, or MetaLog (Marchiori & Saarela, 1998), which represents Horn clauses in RDF(S) format.

The alternative is to fall back to a representation that is even more application specific, viz., the encoding of ontological axioms in pure text, or "CDATA" in RDF speak (cf. the example below). In fact, the latter is a very practical choice for many application-specific axioms—once you make very deep assumptions about a particular representation, you are also free to use whatever format you like.

```
<o:GeneralAxiom rdf:ID="WhoPaidForTheWeddingParty">
<o:text lang="flogic">
    <![CDATA[
    FORALL w, x, y, z
    w:Wedding[groom->x,bride->y,billTo->z] <-
    z[fatherInLawOf->x:Man] AND
    x[marriedWith->y].
    ]]>
</o:text>
</o:GeneralAxiom>
```

Offering such a distinguished place for arbitrary axioms allows *(i)* round tripping of ontologies through different applications, which *(ii)* can benefit as many as possible of these portions of the ontology that are undigestible for others.

RELATED WORK

The proposal described in this paper is based on several related approaches; viz., we have built on considerations made for the RDF inference service SiLRi (Decker et al., 1998), the ontology engineering environments ODE (Blazquez, Fernandez, Garcia-Pinar, & Gomez-Perez, 1998) and Protégé (Grosso, Eriksson, Fergerson, Tu, & Musen, 1999), the ontology interchange language OIL (Horrocks et al., 2000), considerations made by Gruber (Gruber, 1993), and our own earlier work on general ontology engineering (Maedche et al., 2000; Staab & Maedche, 2000).

SiLRi (Decker et al., 1998) was one of the first approaches to propose inferencing facilities for RDF. It provides most of the basic inferencing functions one wants to have in RDF and, hence, has provided a good start for many RDF applications. In fact, it even allows one to use axioms, but these axioms may not be denoted in RDF, but only directly in F-Logic. It lacks capabilities for axiom representation in RDF(S) that our proposal provides.

In our earlier proposals (Maedche et al., 2000; Staab & Maedche, 2000) we have discussed how to push the engineering of ontological axioms from the *symbol level* onto the *knowledge level*—following and extending the general arguments made for ODE (Blazquez et al., 1998) and Ontolingua (Fikes, Farquhar, & Rice, 1997). This strategy has helped us here in providing an RDF(S) object representation for a number of different axiom types.

Nearest to our actual RDF(S)-based ontology engineering tool is Protégé (Grosso et al., 1999), which provides comprehensive support for editing RDFS and RDF. Nevertheless, Protégé currently lacks any support for axiom modeling and inferencing—though our approach may be very easy to transfer to Protégé, too.

A purpose similar to our general goal of representing ontologies in RDF(S) is pursued with OIL (Horrocks et al., 2000). Actually, OIL constitutes an instantiation of our methodological approach, as the definition of concepts and relations in description logics is equivalent to the instantiation of a small number of axiom schemata in a particular logical framework. The axiom categorisation we presented in this paper can be effortlessly combined with the ontological meta-layer proposed in OIL. Thus, applications can utilize two vocabularies that complement each other to model classes in a DL-style while at the same time defining axioms on the conceptual level.

Finally, there are other approaches for ontology exchange and representation in XML formats that we do not want to elaborate here, as they fail our litmus test for supporting the RDF(S) metadata standard (e.g., Karp et al., 1999; Marchiori and Saarela, 1998).

The RDF Schema for Categories of Relationships

```
<?xml version='1.0' encoding='ISO-8859-1'?>
<rdf:RDF
    xmlns:rdf="http://www.w3.org/1999/02/22-rdf-syntax-ns#"
    xmlns:rdfs="http://www.w3.org/TR/1999/PR-rdf-schema-
    19990303#">
<rdfs:Class ID="Relation">
<rdfs:subClassOf rdf:resource="http://www.w3.org/1999/02/22-
    rdf-syntax-ns#Property"/>
</rdfs:Class>
<rdfs:Class ID="Asymmetric">
<rdfs:subClassOf rdf:resource="#Relation"/>
</rdfs:Class>
<rdfs:Class ID="Reflexive">
<rdfs:subClassOf rdf:resource="#Relation"/>
</rdfs:Class>
<rdfs:Class ID="Transitive">
<rdfs:subClassOf rdf:resource="#Relation"/>
</rdfs:Class>
<rdfs:Class ID="Irreflexive">
<rdfs:subClassOf rdf:resource="#Relation"/>
</rdfs:Class>
<rdfs:Class ID="Symmetric">
<rdfs:subClassOf rdf:resource="#Relation"/>
</rdfs:Class>
<rdfs:Class ID="PartOfRel">
<rdfs:subClassOf rdf:resource="#Relation"/>
</rdfs:Class>
<rdf:Description ID="isInverseRelationOf">
<rdf:type rdf:resource="#Relation"/>
</rdf:Description>
<!-- Definitions for COMPOSITION -->
<rdfs:Class ID="Composition"/>
<rdf:Property ID="composee">
<rdfs:domain rdf:resource="#Composition"/>
<rdfs:range  rdf:resource="http://www.w3.org/1999/02/22-rdf-
    syntax-ns#Property"/>
</rdf:Property>
<rdf:Property ID="firstComponent">
<rdfs:domain rdf:resource="#Composition"/>
```

```
<rdfs:range rdf:resource="http://www.w3.org/1999/02/22-rdf-
    syntax-ns#Property"/>
</rdf:Property>
<rdf:Property ID="secondComponent">
<rdfs:domain rdf:resource="#Composition"/>
<rdfs:range rdf:resource="http://www.w3.org/1999/02/22-rdf-
    syntax-ns#Property"/>
</rdf:Property>
<!-- Definitions for PARTITION -->
<rdfs:Class ID="Partition"/>
<rdf:Property ID="partitionee">
<rdfs:domain rdf:resource="#Partition"/>
<rdfs:range rdf:resource="http://www.w3.org/1999/02/22-rdf-
    syntax-ns#Property"/>
</rdf:Property>
<rdf:Property ID="parts">
<rdfs:domain rdf:resource="#Partition"/>
<rdfs:range rdf:resource="http://www.w3.org/1999/02/22-rdf-
    syntax-ns#Bag"/>
</rdf:Property>
<rdfs:Class ID="PartonomicRolePropagation"/>
<!-- Definitions for General Axioms-->
<rdfs:Class ID="GeneralAxiom">
<rdfs:subClassOf rdf:resource="http://www.w3.org/TR/1999/PR-
    rdf-schema-19990303#Resource"/>
</rdfs:Class>
<rdf:Property ID="lang">
<rdfs:domain rdf:resource="GeneralAxiom"/>
<rdfs:range rdf:resource="http://www.w3.org/TR/xmlschema-2/
    #string"/>
</rdf:Property>
<rdf:Property ID="text">
<rdfs:domain rdf:resource="GeneralAxiom"/>
<rdfs:range rdf:resource="http://www.w3.org/TR/xmlschema-2/
    #string"/>
</rdf:Property>
</rdf:RDF>
```

DISCUSSION

We have presented a new approach towards engineering ontologies extending the general arguments made for ODE (Blazquez et al., 1998) and Ontolingua (Fikes et al., 1997) in the Web formats, RDF and RDFS. Our objectives aim at the usage of existing inferencing services such as provided by deductive database mechanisms (Decker et al., 1998) or description logics systems (Horrocks, 1998). We reach these objectives through a *methodology* that classifies axioms into axiom types according to their *semantic meaning*. Each type receives an object representation that abstracts from scoping issues and is easily representable in RDF(S). Axiom descriptions only keep references to concepts and relations necessary to distinguish one particular axiom of one type from another one of the same type. When the limits of object representations in RDF(S) are reached, we fall back onto target system-specific representations. These may be formulated in RDF versions of languages like OIL or MetaLog—but since they are commonly very specific for particular applications, they may also be expressed by strings (CDATA), the particular semantics of which is only defined in the corresponding application.

Our proposed extension of RDF(S) has been made with a clear goal in mind—the complete retention of the expressibility and semantics of RDF(S) for the representation of ontologies. This includes the relationship between ontologies and instances, both represented in RDF(S). Especially, the notion of *consistency* (cf. Brickley & Guha, 1999) between an RDF model and a schema also holds for ontologies expressed in RDF(S). The integration of the newly defined resources has been carried out in such a way that all RDF processors capable of processing RDF schemas can correctly interpret RDF models following the ontology schema, even if they do not *understand* the semantics of the resources in the o-namespace.

Special applications like OntoEdit (Maedche et al., 2000) can interpret the o-namespace correctly and thus fully benefit from the richer modeling primitives, if the RDF model is valid (cf. the "Validator" section in http://www.ics.forth.gr/proj/isst/RDF/ for a set of operations to check for validity) according to the defined ontology schema. Our approach has been partially implemented in our ontology engineering environment, OntoEdit. The object-model engineering capabilities for RDF(S) are ready to use, while different views for axiom representations are currently under construction.

ACKNOWLEDGEMENTS

The research presented in this chapter has been partially funded by BMBF under grant number 01IN802 (project "GETESS"). We thank our student Dirk Wenke, who implemented large parts of the RDF(S)-based ontology editor.

REFERENCES

Blazquez, M., Fernandez, M., Garcia-Pinar, J. M. and Gomez-Perez, A. (1998). Building ontologies at the knowledge level using the ontology design environment. In *Proceedings of the 11th Int. Workshop on Knowledge Acquisition, Modeling and Mangement (Kaw'98),* Banff, Canada, October.

Brickley, D. and Guha, R. (1999). *Resource Description Framework (RDF)*

Schema Specification (Tech. Rep.). *W3C. W3C Proposed Recommendation*. Retrieved on the World Wide Web: http://www.w3.org/TR/PR-rdf-schema/.

Decker, S. (1998). On domain-specific declarative knowledge representation and database languages. In *Proceedings of the 5th Knowledge Representation Meets Databases Workshop (Krdb'98)*.

Decker, S., Brickley, D., Saarela, J. and Angele, J. (1998). A query and inference service for RDF. In *Ql'98–The Query Languages Workshop*. W3C. Retrieved on the World Wide Web: http://www.w3.org/TandS/QL/QL98/.

Fikes, R., Farquhar, A. and Rice, J. (1997). Tools for assembling modular ontologies in Ontolingua. In *Proceedings of AAAI 97*, 436-441.

Grosso, E., Eriksson, H., Fergerson, R. W., Tu, S. W. and Musen, M. M. (1999). Knowledge modeling at the millennium—The design and evolution of Protégé-2000. In *Proceedings of the 12th International Workshop on Knowledge Acquisition, Modeling and Management (Kaw'99)*, Banff, Canada, October.

Gruber, T. (1993). A translation approach to portable ontology specifications. *Knowledge Acquisition*, 5(1), 199-220.

Horrocks, I. (1998). Using an expressive description logic: FaCT or fiction? In *Proceedings of KR-98*, 636-647.

Horrocks, I., Fensel, D., Broekstra, J., Decker, S., Erdmann, M., Goble, C., Harmelen, F. V., Klein, M., Staab, S. and Studer, R. (2000). *The Ontology Interchange Language Oil: The Grease Between Ontologies* (Tech. Rep.). Department of Computer Science, University of Manchester, UK/ Vrije Universiteit Amsterdam, NL/ AIdministrator, Nederland B.V./ AIFB, University of Karlsruhe, DE. Retrieved on the World Wide Web: http://www.cs.vu.nl/~dieter/oil/.

Karp, P. D., Chaudhri, V. K. and Thomere, J. (1999). *XOL: An XML-Based Ontology Exchange Language* (Tech. Rep.) (Version 0.3).

Kifer, M., Lausen, G. and Wu, J. (1995). Logical foundations of object-oriented and frame-based languages. *Journal of the ACM*, 42.

Lassila, O. and Swick, R. R. (1999). *Resource Description Framework (RDF) Model and Syntax Specification* (Tech. Rep.). W3C. *W3C Recommendation*. Retrieved on the World Wide Web: http://www.w3.org/TR/REC-rdf-syntax.

Maedche, A., Schnurr, H.-P., Staab, S. and Studer, R. (2000). Representation language-neutral modeling of ontologies. In Frank, U. (Ed.), *Proceedings of the German Workshop "Modellierung-2000."* Koblenz, Germany, April, 5-7, Fölbach-Verlag.

Marchiori, M. and Saarela, J. (1998). Query + metadata + logic = metalog. In *Ql '98-The Query Languages Workshop*. W3C. Retrieved on the World Wide Web: http://www.w3.org/TandS/QL/QL98/.

Staab, S. and Maedche, A. (2000). *Axioms are Objects, Too—Ontology Engineering Beyond the Modeling of Concepts and Relations* (Tech. Rep. No. 399). Institute AIFB, University of Karlsruhe.

Weibel, S. and Miller, E. (1998). *Dublin Core Metadata* (Tech. Rep.). Retrieved on the World Wide Web: http://purl.oclc.org/dc.

Chapter XIV

Towards a Semantic Web of Evidence-Based Medical Information

Rolf Grütter and Claus Eikemeier
University of St. Gallen, Switzerland

Johann Steurer
University Hospital of Zurich, Switzerland

INTRODUCTION AND OBJECTIVES

It is the vision of the protagonists of the Semantic Web to achieve "a set of connected applications for data on the Web in such a way as to form a consistent logical Web of data" (Berners-Lee, 1998, p. 1). Therefore, the Semantic Web approach develops languages for expressing information in a machine-processable form ("machine-understandable" in terms of the Semantic Web community). Particularly, the Resource Description Framework, RDF (Lassila & Swick, 1999), and RDF Schema, RDFS (Brickley & Guha, 2000), are considered as the foundations for the implementation of the Semantic Web. RDF provides a data model and a serialization language; RDFS a distinguished vocabulary to model class and property hierarchies and other basic schema primitives that can be referred to from RDF models, thereby allowing for the modeling of object models with cleanly defined semantics. The idea behind this approach is to provide a common minimal framework for the description of Web resources while allowing for application-specific extensions (Berners-Lee, 1998). Such extensions in terms of additional classes and/or properties must be documented in an application-specific schema. Application-specific schemata can be integrated into RDFS by the namespace mechanism (Bray, Hollander & Layman, 1999). Namespaces provide a simple method for qualifying element and attribute names used in RDF documents by associating them with namespaces identified by URI (Uniform Resource Identifier) references (Berners-Lee, Fielding, Irvine & Masinter, 1998).

With the objective to facilitate integration of information from distributed and heterogeneous sources, this chapter describes the integration of an existing Web-based ontology on evidence-based medicine into the RDF/RDFS framework. Following the presentation of the application context, the applied conceptual framework will be introduced. The main part of the chapter will then describe the modeling of the ontology with RDF/RDFS. This will provide the basis for a planned re-implementation of the existing ontology. A discussion of the presented approach will conclude the chapter. The contribution of this chapter rests upon the application of methods which are to some extent already established to a real-world scenario. Based on this application, the scope of the term "ontology" within the RDF/RDFS framework will be redefined, particularly by introducing a Simple Ontology Definition Language (SOntoDL). This re-definition contributes to the implementation of the Semantic Web and to ontology modeling in general.

APPLICATION CONTEXT: THE EVIMED PROJECT

The term "evidence-based medicine" was coined at the McMaster Medical School, Canada, in the eighties. According to Sackett, a co-founder of evidence-based medicine (EBM), EBM is the conscious, explicit, and judicious application of the currently best evidence in making medical care decisions in favor of an individual patient (Sackett, Rosenberg, Gray, Haynes & Richardson, 1996). Practicing EBM means integrating the personal clinical experience and the best available external clinical evidence, which can be derived from systematic analyses of published studies. The key issues differentiating EBM from other approaches towards providing high quality and relevant medical information are the significant contribution of humans adding value in terms of expert knowledge, particularly to downstream activities, such as judging the methods applied in a particular study, and the tailoring of the information according to the needs of the consultation hour.

Evidence-based medical information services are quite new on the Internet (Hersh, 1996). For instance, the Cochrane Collaboration offers free abstracts of reviews by almost 50 specialized collaborative review groups (The Cochrane Library Issue 3, 2000). In order to access the full text of these reviews, subscription is required. Recently, Ovid Technologies (2000) has introduced a similar evidence-based medicine review service. Ovid integrates (1) the Cochrane database of systematic reviews, (2) Best Evidence, a database that contains the *ACP Journal Club* (a publication of the American College of Physicians–American Society of Internal Medicine), and *Evidence-Based Medicine* (a joint publication of the ACP and British Medical Journal Group), and (3) the database of abstracts of reviews of effectiveness produced by the expert reviewers and information staff of the National Health Service's Centre for Reviews and Dissemination (NHS CRD). Ovid includes a natural language processor for the interpretation of free-text queries in English. On-line access to the service requires a proprietary, purchasable software. Alternatively, contents are available on CD-ROM.

The Evimed project (www.evimed.ch) is a similar evidence-based medical information service on the Internet. It was initiated in April 1998 and aims at providing general practitioners in the German-speaking countries with relevant and reliable information for daily practice, thereby supporting the practitioners in making appropriate medical decisions. To achieve this, a group of physicians who are trained in evidence-based medicine systematically reviews published studies with respect to practical relevance and trustworthiness. These reviews are published together with the links to the original articles in the Journal Club of Evimed (Figure 1).

The Journal Club is the core of the Evimed Web site. Currently (i.e., in November 2000) it stores 330 reviews of selected articles published in various biomedical journals. The reviews are categorized according to 21 specialties and can either be browsed or be accessed via a (syntactic) search engine. The Web site includes a separate section with articles on the subject of EBM. Evimed also includes a glossary with currently 25 definitions of EBM-specific concepts. Further services include access to Medline, links to literature and other EBM-related sources (e.g., a calendar of events related to further education in EBM), and a guest book offering the possibility to post comments and to subscribe to a newsletter. Taking into account the objection of practitioners who do not feel comfortable with translating foreign languages, all text on the Web site is in German. (This is a basic difference to the mentioned Cochrane Collaboration and Ovid, which offer their contents in English only.) To date, all services of Evimed are freely available.

Figure 1: Journal Club of the Evimed Web site

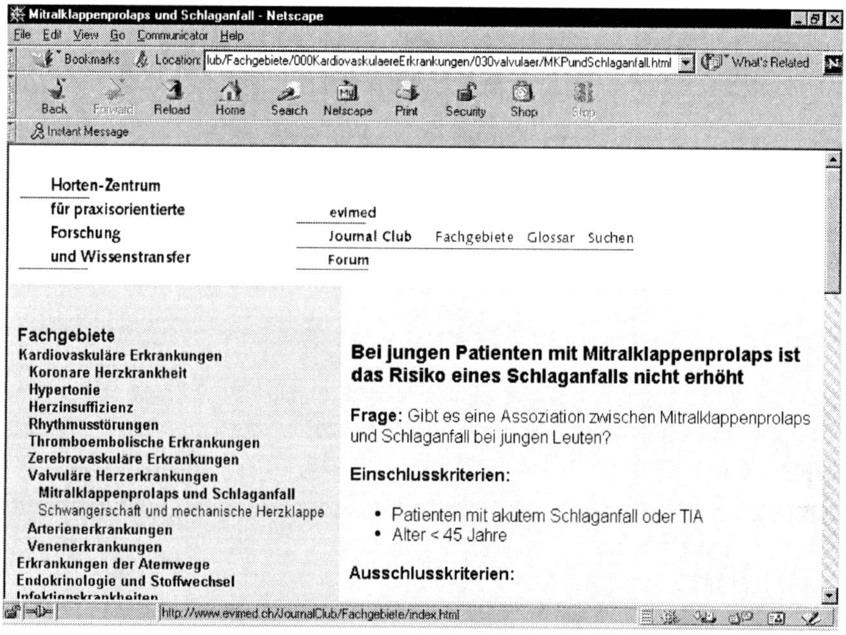

METHODS: THE CONCEPTUAL FRAMEWORK OF THE ONTOLOGY

The applied conceptual framework refers to the concept of the ontology as defined by Gruber (2000). According to Gruber, an ontology is a *specification of a conceptualization*, i.e., a formal description of the concepts and their relations for a "universe of discourse." The universe of discourse refers to the set of objects which can be represented in order to represent the (propositional) knowledge of a domain. This set of objects and the describable relations among them are reflected in a representational vocabulary. In an ontology, definitions associate the names of objects in the universe of discourse with human-readable text, describing what the names mean, and formal axioms constrain the interpretation and well-formed use of the ontology. In short, an ontology consists of the triple (vocabulary, definitions, axioms). Formally, an ontology is the statement of a logical theory.

A prototype implementation of an ontology supporting a semantic navigation of the Evimed Web site is described by Grütter, Eikemeier, and Steurer (2001). Thereby, EBM-specific concepts in the reviews of the Journal Club (e.g., "Sensitivität") are linked via JavaScript function calls to the inference engine which retrieves the definition of the requested concept together with options for a further semantic navigation. Based on the experience with this prototype, the current implementation is reconsidered for the following reasons.

(1) The prototype implementation takes advantage of the intrinsic structuring capabilities of XML (Extensible Markup Language) documents, i.e., the representational vocabulary is implemented as a hierarchy of items together with their identifiers and descriptions, whereby each item corresponds to a node of the document tree. This pragmatic approach has been chosen since the Evimed vocabulary was provided as a hierarchy of concepts. It yields the advantage that the types of relation between the concepts must not be made explicit (i.e., `parentOf`, `siblingOf`, `childOf`). Instead, the type `childOf` is accessed from the Document Object Model (DOM, i.e., the Application Programming Interface, API) after parsing and, based thereon, the remaining types are reconstructed by the inference engine. The disadvantage of this approach is its limitation to a hierarchical representation and the inability to represent more complex knowledge structures. (Basically, the attribute types ID and IDREF of the XML syntax specification (Bray, Paoli, Sperberg-McQueen & Maler, 2000) allow also for an nonhierarchical representation. However, since such a representation has to be defined on the document level (by instantiating ID and IDREF), a detailed analysis of the document by the inference engine is required which far exceeds the implemented identification of its general structure based on the document type definition (DTD). This kind of analysis cannot easily be implemented by the inference engine without jeopardizing the targeted independence of the components.)

(2) The representational vocabulary of the prototype implementation also includes the human-readable definitions. As a disadvantage, the XML document is relatively large, not in terms of file size but of number of screens to scroll in order to edit the vocabulary (unless an XML editor is used instead of a text editor). This point will become even more important in the future, as the vocabulary will grow in accordance with the evolving knowledge base of the domain. Moreover, updates of existing definitions must be made in this central document and cannot be easily distributed to decentralized editors, thereby preventing a physically dissociated maintenance of representational vocabulary and human-readable definitions.

(3) The prototype ontology is not yet integrated into a common framework which allows for an easy integration of information from distributed and heterogeneous sources.

These shortcomings result in the following requirements for a redefinition of the Evimed ontology:

(1) The vocabulary must allow for the representation of more complex knowledge structures than hierarchical ones only.

(2) The representational vocabulary and the human-readable definitions must be physically dissociated.

(3) The ontology must be integrated into a framework which has the potential to become the de facto standard for interconnected applications on the Web, thereby taking advantage of a range of tools being developed.

As mentioned in the Introduction, the Semantic Web initiative promotes RDF/RDFS as a basic framework. Therefore, this chapter describes the re-definition of the existing Evimed ontology using RDF/RDFS. The re-definition will be described in three steps:

(1) Directed labeled graph (DLG) representation of a sample item of the ontology (i.e., a resource in terms of RDF/RDFS) and its serialization in RDF syntax.

(2) Definition of an application-specific schema using RDFS and its serialization in RDF syntax. Its objective is to provide the semantics of the application-specific concepts and their relations introduced by the DLG representation. The application-specific schema is referred to by the newly introduced `sontodl` namespace. It provides the classes of an ontology definition language.

(3) Serialization of the ontology in RDF syntax based on the specified application-specific schema, i.e., application of the ontology definition language to the Evimed domain.

In order to make the process more explicit, the second step will be further divided into two sub-steps: (1) definition of the rough schema using the `rdf` and `rdfs` namespaces, and (2) detailing of the schema using an additional namespace specified by Staab, Erdmann, Maedche, and Decker (2000).

RESULTS: DEFINITION OF AN ONTOLOGY FOR EVIMED

Directed Labeled Graph Representation

Figure 2 shows a DLG representation of a sample item of the Evimed ontology. In this figure, the nodes (drawn as ovals) represent resources and arcs represent named properties. Nodes that represent string literals are drawn as rectangles. The DLG representation comprises five statements:

(1)   The resource http://www.evimed.ch/JournalClub/Glossar/RRR.html has the
       label "Relative Risikoreduktion."

(2)   The resource http://www.evimed.ch/JournalClub/Glossar/RRR.html has
       the comment "This is the definition of the labeled concept by Evimed."

(3)   The resource http://www.evimed.ch/JournalClub/Glossar/RRR.html is the
       child of the resource http://www.evimed.ch/JournalClub/Glossar/CER.html.

(4)   The resource http://www.evimed.ch/JournalClub/Glossar/RRR.html is the
       sibling of the resource http://www.evimed.ch/JournalClub/Glossar/ARR.html.

(5)   The resource http://www.evimed.ch/JournalClub/Glossar/RRR.html is the
       parent of the resource http://www.evimed.ch/JournalClub/Glossar/NNT.html.

The serialization of the DLG representation in RDF syntax results in the
following description (note that the serialization is not ready for the processing by an XML parser (RDF is an XML application); for a complete
serialization cf. Appendix II).

```
<rdf:RDF>
  <rdf:Description about="http://www.evimed.ch/JournalClub/
  glossar/RRR.html">
    <rdfs:label xml:lang="de">Relative Risikoreduktion
    </rdfs:label>
    <rdfs:comment xml:lang="en">This is the definition of the
    labeled concept by Evimed.</rdfs:comment>
    <sontodl:childOf  resource="http://www.evimed.ch/
    JournalClub/Glossar/CER.html"/>
    <sontodl:siblingOf  resource="http://www.evimed.ch/
    JournalClub/Glossar/ARR.html"/>
    <sontodl:parentOf  resource="http://www.evimed.ch/
    JournalClub/Glossar/NNT.html"/>
  </rdf:Description>
</rdf:RDF>
```

From a knowledge engineering point of view, the DLG of Figure 2
represents a labeled concept together with pointers to its definition and to
related concepts whereby the relations `sontodl:childOf` and
`sontodl:parentOf` point to sub- and superordinated concepts, and the
relation `sontodl:siblingOf` to a concept at the same level. If all concepts
of the vocabulary are represented this way, a semantic network is created
which represents in its entirety the knowledge of the Evimed domain.
Therefore, the RDF model can be regarded as the knowledge representation
layer of the ontology.

    In addition to the representation as implemented in the existing ontology, the
DLG allows an item to have multiple parents and supports further types of relation
(e.g., `sontodl:partOf`, not shown). The first allows for the representation of
more complex knowledge structures than simple hierarchies; the second, among

Figure 2: Directed labeled graph representation of an Evimed ontology item

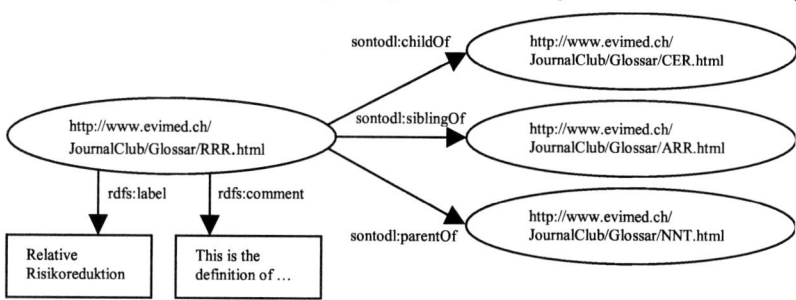

others, for a clear distinction between generic and partitive relations, which is a precondition for the logic-based processing of medical terminologies (Schulz, Romacker & Hahn, 1998).

Application-Specific Schema

Figure 3 shows step (1) of the definition of an application-specific schema for the introduced RDF model and its integration into the RDF Schema. The schema layers are drawn in bold face. Note that for a better readability not all classes and properties which are part of the RDF/RDFS layer and namespace are drawn (e.g., the property `rdfs:label` is omitted).

The application-specific schema is referred to by the newly introduced `sontodl` namespace. It provides the classes of a Simple Ontology Definition Language (SOntoDL; cf. Discussion).

Although the schema allows one to assign the introduced properties to the items by the `rdfs:domain` property (not shown) and to constrain the range of valid values by the `rdfs:range` property (not shown), it does not specify the *types* of relations applied in the DLG representation. In order to achieve this, an approach described by Staab et al. (2000) is applied. It builds on the modeling of the semantics of types of relations by assigning these to classes of axioms which are defined as subclasses of `rdf:Property`. The DLG of Figure 2 represents a set of triples of type (resource, named property, property value) or, more generally, (subject, predicate, object). In other words, the RDF model defines a set of binary relations. Therefore, semantic consistency should be describable by the axioms of the relational algebra. These include reflexivity, irreflexivity, symmetry, asymmetry, antisymmetry, transitivity, and intransitivity. The axioms for a relational algebra are subsequently formally described using a first-order predicate logic.
- Reflexivity: $\forall$ x R(x, x)
  Description: For any element x holds the relation R.

Figure 3: Integration of directed labeled graph representation into RDF Schema

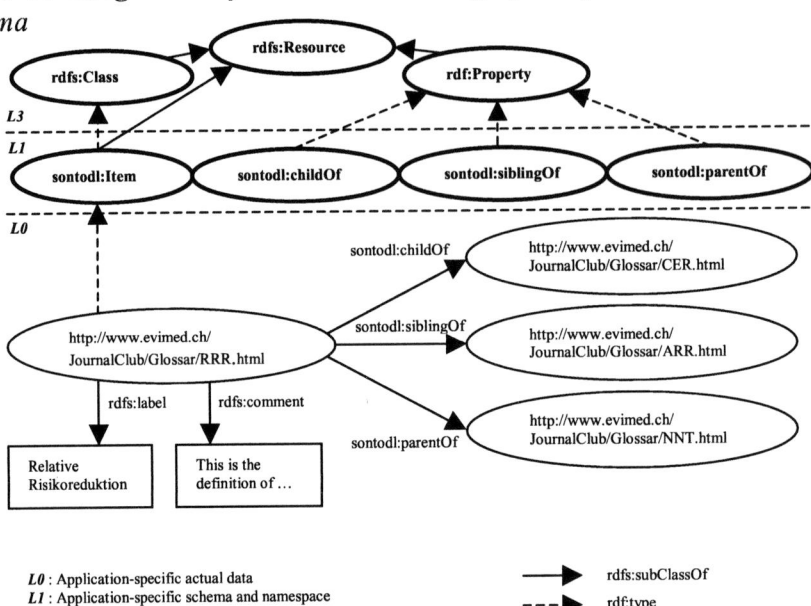

- Irreflexivity: $\neg \exists x \, R(x,x)$
  Description: There is no element x for which reflexivity holds.
- Symmetry: $\forall x, y \, R(x, y) \rightarrow R(y, x)$
  Description: If for any pair (x, y) the relation R holds, then also its inverse R^{-1} holds.
- Asymmetry: $\neg \exists x, y \, R(x, y) \rightarrow R(y, x)$
  Description: There is no pair (x, y) for which symmetry holds.
- Antisymmetry: $\forall x, y \, R(x, y) \wedge R(y, x) \rightarrow x = y$
  Description: If for any pair (x, y) both the relation R and its inverse R^{-1} hold, then x is identical with y.
- Transitivity: $\forall x, y, z \, R(x, y) \wedge R(y, z) \rightarrow R(x, z)$
  Description: If for any two pairs (x, y) and (y, z) the relation R holds, then R also holds for the pair (x, z).
- Intransitivity: $\neg \exists x, y, z \, R(x, y) \wedge R(y, z) \rightarrow R(x, z)$
  Description: There are no two pairs (x, y) and (y, z) for which transitivity holds.

Table 1 describes the types of relation of the ontology by assigning the valid axioms. It shows that the axioms for a relational algebra are not sufficient for an unambiguous description. In particular, the properties sontodl:parentOf and sontodl:childOf would have the same descrip-

Table 1: Types of relation and valid axioms

Type of relation	Axioms
`sontodl:parentOf`	irreflexivity, asymmetry, intransitivity
`sontodl:siblingOf`	irreflexivity, symmetry, transitivity
`sontodl:childOf`	irreflexivity, asymmetry, intransitivity, inverse relation

tion, although they are not identical. Therefore, the types of relation are further qualified according to their interrelationship: `sontodl:childOf` is the inverse relation of `sontodl:parentOf`.

Inverse relation: $\forall$ x, y R(x, y) $\rightarrow$ R^{-1}(y, x)

Description: If for any pair (x, y) the relation R holds, then its inverse R^{-1} holds for any pair (y, x).

(Note that in RDF/RDFS it is prohibited that any class is an `rdfs:subClassOf` itself or of one of its subclasses. Therefore, a generic relation between two items, as implied by the property `sontodl:childOf`, cannot be modeled by an `rdfs:subClassOf` property.)

From the axioms for a relational algebra the following integrity constraints for the Evimed ontology can directly be derived:

(1) If the resource R_i is a child of the resource R_k, then R_k is the parent of R_i (i, k $\in$ N, i $\neq$ k).

(2) If the resource R_i is the parent of the resource R_k, then R_k is a child of R_i (i, k $\in$ N, i $\neq$ k).

(3) If the resource R_i is the sibling of the resource R_k, then R_k is the sibling of R_i (i, k $\in$ N, i $\neq$ k).

Figure 4 shows step (2) of the definition of an application-specific schema. It introduces an additional ontology meta-layer and namespace. By defining the introduced properties in the DLG representation of Figure 2 as instances of the applicable classes of axioms, these are semantically described. Note that for a better readability not all classes of axioms are drawn.

The application-specific schema can be serialized in RDF syntax as shown in Appendix I. Based on the definition of the application-specific schema and on additional schemata, which are referred to by the namespaces `dc` and `odoc`, the sample item of the Evimed ontology of Figure 2 can be serialized in RDF syntax as shown in Appendix II.

Figure 4: Detailing of application-specific schema

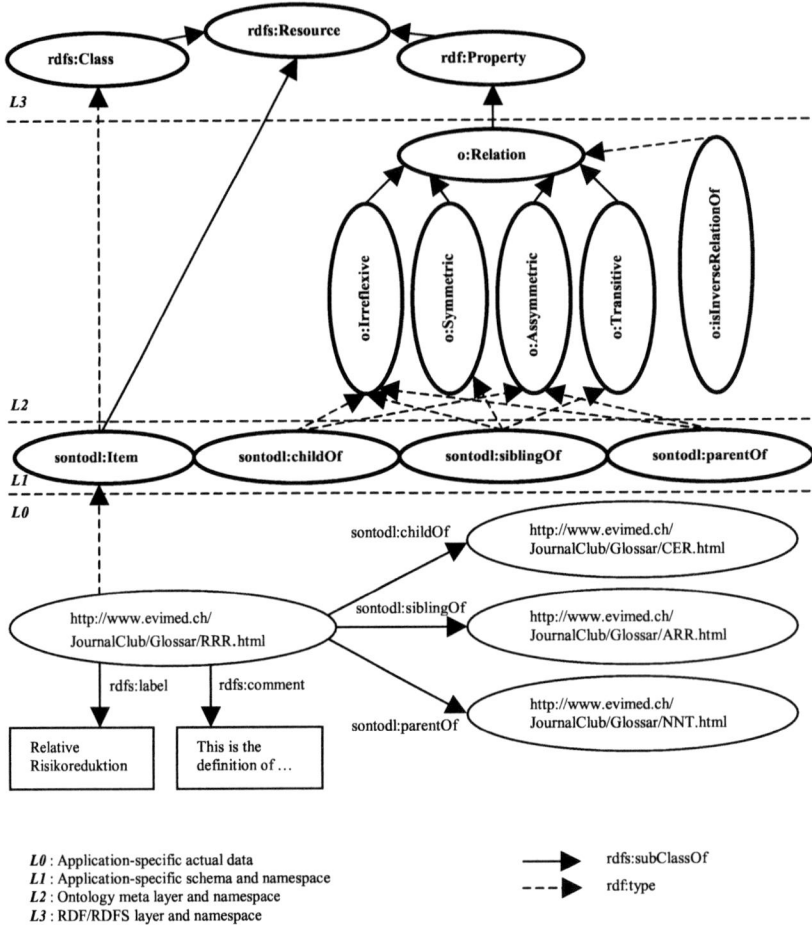

DISCUSSION

Although the approach to modeling an ontology for an evidence-based medical information service on the Web is borrowed from Staab et al. (2000), there is a basic difference in the scope of the term "ontology." Whereas for Staab et al. (2000) the schema to which the application-specific namespace refers *is* the ontology, in the presented approach the ontology comprises the resources, the RDF descriptions, and the schema. Particularly, the resources provide the human-readable definitions of the conceptual framework introduced in the Methods section, the RDF descriptions implement the representational vocabulary, and the schema represents the axioms in terms of integrity constraints. If, different from the presented application, the resources and the human-readable definitons are not the same, the latter can be implemented as part of the RDF descriptions, e.g., by using the `rdfs:comment` property (cf. Figure 2). Taken alone, the schema of the Evimed ontology specifies an ontology *definition language* rather than a particular ontology. The reason for this conception is the intended generalization of the approach to any kind of domain.

The specified ontology definition language, i.e., SOntoDL, allows Evimed to easily integrate external resources (by simple URI references). Conversely, if an editor of another Web site wants to refer to a definition as introduced by Evimed he/she can do so by a simple hyperlink. Whereas the latter does not work with the prototype ontology (the definitions are embedded in the representational vocabulary), both cases do not necessitate an integration into the RDF/RDFS framework. Integration into RDF/RDFS is only required if special tools such as search engines running over RDF descriptions should be supported. Given the huge number of Web sites and the unsatisfactory precision and completeness of the common syntactic search engines, these RDF search tools can add a lot of value by finding information which is *semantically* related. Therefore, it is reasonable to integrate the Evimed ontology into RDF/RDFS. After all, the subject of Berners-Lee's vision (1998) is integration: "... sometimes it is less than evident why one should bother to map an application in RDF. The answer is that we expect this data, while limited and simple within an application, to be combined, later, with data from other applications into a Web. Applications which run over the whole Web must be able to use a common framework for combining information from all these applications" (p. 2). Additional applications Evimed will probably take advantage of include RDF editors. An easy maintenance of the representational vocabulary is a requirement from the user's point of view (i.e., the healthcare professionals of Evimed), and RDF editors can be evaluated for their applicability as an ontology editor, thereby avoiding unnecessary programming.

There are several ways how RDF descriptions can be associated with the resource they describe (Lassila & Swick, 1999):

(1) The description may be contained within the resource ("embedded"; e.g., in HTML).

(2) The description may be external to the resource but supplied by the transfer mechanism in the same retrieval transaction as that which returns the resource ("along-with"; e.g., with HTTP GET or HEAD).

(3) The description may be retrieved independently from the resource, including from a different source ("service bureau"; e.g., using HTTP GET).

(4) The description may contain the resource ("wrapped"; e.g., RDF itself).

With respect to the Evimed ontology, the option (3) is considered, where the description is retrieved independently from the resource. A repository for the descriptions together with another repository storing the application-specific schema still has to be established (this is the reason why dummy URIs are used for the localization of the repositories in the serialization of the sample ontology item in Appendix II).

CONCLUSION AND OUTLOOK

With the prototype ontology, an ontology definition language was defined by a rather generic DTD (Grütter et al., 2001), and based on this language the representational vocabulary was implemented as an XML document. From a software engineering point of view, the DTD can be regarded as the specification for the inference engine. Therefore, this generic approach allowed the engine to reason over a range of vocabularies from different domains. The only requirement was that their representations comply with the DTD (i.e., that they are valid XML documents). In that way, the requirement of generality (and of independence of components) was met. With the redefinition of the representational vocabulary as presented in this chapter, the DTD of the prototype is mapped to the application-specific schema, resulting in an additional four classes (i.e., `sontodl:Item`, `sontodl:childOf`, `sontodl:siblingOf`, `sontodl:parentOf`). These are instantiated in the DLG representation and serialized in the RDF syntax as element types and hence will appear in a hypothetical RDF DTD as element type declarations (not shown). Since the RDF DTD contains additional declarations for the element types and attribute lists of the serialization (or abbreviated) syntax, the specification for the inference engine has changed and a reprogramming is required (although the RDF DTD and RDF Schema are almost interchangeable concepts, the lean EBNF-like notation of the DTD is better suited as a specification syntax than the somewhat extensive RDF serialization of the schema). However, before reprogramming the inference engine, existing tools should be evaluated with respect to their inferencing capabilities. A list of such tools can be found on the RDF Web site (http://www.w3.org/RDF/).

Another approach to avoid reprogramming of the inference engine is the preprocessing of the RDF representation using XSLT (Clark, 1999), particularly the transformation of RDF into the representation of the prototype ontology. This could be a reasonable solution, since the representational vocabulary of the prototype ontology is parsed only once for a session, and all subsequent user interaction is handled by a number of nonpersistent JavaScript objects in the random-access memory (RAM) of the client computer. However, due to the hierarchical nature of the prototype representation preprocessing would not allow for the transformation of more complex knowledge structures.

The question remains whether or not RDF/RDFS will succeed as the technology of the Semantic Web. A competitor, i.e., the Topic Map standard (Biezunski, Bryan & Newcomb, 1999) has recently been issued. This standard defines an SGML/XML architectural syntax for the representation of topics and semantic relations between topics. The Topic Map architecture has particularly been developed with the aim to provide access to indices, glossaries, thesauri, etc., based on a model of the knowledge they contain. It seems that the topic paradigm has a broad field of applications and can serve as a general basis for navigating information. To our knowledge, currently (i.e., in December 2000), there is a single tool available that supports the standard, i.e., a "Topic Map Navigator" (Ontopia, 2000). The future will show which standard succeeds as the key technology to approach the vision of the Semantic Web.

RELATED WORK

SHOE (Simple HTML Ontology Extensions) provides distributed on-tologies consisting of categories and relationship rules (Heflin, Hendler, & Luke, 1999). Thereby, the categories provide for the classification of in-stances. They are organized hierarchically and support multiple inheritance. The relationship rules are implemented as Horn clauses. The instances (i.e., individual constants in terms of Horn rules) are represented as URLs/URIs. This is similar to the approach as presented in this chapter, where the human-readable definitions, as part of the ontology, are likewise represented by URI references. SHOE was originally specified in SGML (as is HTML; before the definition of XML) but is meanwhile also specified as an XML DTD.

XOL (XML-based Ontology Exchange Language) is a language for specifying and exchanging ontologies (Karp, Chaudhri, & Thomere, 1999). XOL is specified in an XML-based syntax (kernel DTD). It uses a frame-based semantic model, i.e., OKBC-Lite. An XOL file consists of a module-header definition and one or more class, slot and individual definitions. The module-header definition provides meta-information of the ontology, such as the name and version. The class definitions provide the classes and subclasses of the defined individuals. The slot definitions are strings that encode the official names of the entities. Each slot definition refers to a class name. The individual definitions provide the names, documentations, instance-of infor-mation, and slot-values of the defined individuals. As a disadvantage, XOL does not reuse the core semantics of RDF/RDFS. Hence, "pure" RDF/RDFS applications cannot process even the core object-model definitions.

The Ontobroker application answers queries based on a facts base and an ontology base (Decker, Erdmann, Fensel, & Studer, 1999). The facts base stores instance information („values" in terms of the query interface) which is extracted from annotated HTML pages, for instance, of a corporate Intranet, which is different from the hereby presented approach, where the HTML pages, i.e., the resources, are externally annotated by RDF descriptions. The ontology base stores a set of ontologies. Each ontology includes a concept hierarchy ("classes"), a set of slot definitions ("attributes"), and a set of rules. The rules implement integrity constraints for the ontology. Ontologies are defined as F-Logic statements.

An approach to representing ontologies in RDF/RDFS, similar to Staab et al. (2000; and to the approach as presented in this chapter) is pursued with OIL (Ontology Interchange Language; Horrocks et al., 2000). OIL uses description logics for the definition of concepts and relations and proposes an ontological meta-layer that is combinable with the herein applied axiom categorization proposed by Staab et al. (2000).

Closely related is the recent approach as pursued by DAML (DARPA Agent Markup Language; Hendler, 2000). The goal of the DAML program is to create technologies that enable software agents to dynamically identify and understand information sources, and to provide interoperability between agents in a semantic manner. Particularly, an agent markup language developed as an extension to XML and RDF should allow users to provide machine-readable semantic annotations for specific communities of interest. According to the initiators of the DAML program, and similar to Ontobroker, objects in the Web will be marked to include descriptions of information they encode, of functions they provide, and/ or of data they can produce. In addition, DAML should allow for an "ontology calculus" similar to the relational calculus that makes DataBase Management Systems (DBMS) possible.

ACKNOWLEDGMENTS

The authors would like to thank Andrea Kaiser from the Institute for Media and Communications Management, University of St. Gallen, Switzerland for the carefully drawn figures to this chapter. Likewise, the thorough review and the helpful suggestions of Markus Greunz from the same institute are gratefully acknowledged.

REFERENCES

Berners-Lee, T. (1998). *Semantic Web Road map*. Retrieved December 28, 2000 on the World Wide Web: http://www.w3.org/DesignIssues/Semantic.html.

Berners-Lee, T., Fielding, R., Irvine, U. C. and Masinter, L. (1998). *Uniform Resource Locators (URI): Generic Syntax*. Retrieved October 16, 2000 on the World Wide Web: http://www.ietf.org/rfc/rfc2396.txt.

Biezunski, M., Bryan, M. and Newcomb, S. R. (1999). *ISO/IEC FCD 13250:1999–Topic Maps*. Retrieved December 27, 2000 on the World Wide Web: http://www.ornl.gov/sgml/sc34/document/0058.htm.

Bray, T., Hollander, D. and Layman, A. (1999). *Namespaces in XML. World Wide Web Consortium 14-January-1999*. Retrieved July 26, 2000 on the World Wide Web: http://www.w3.org/TR/1999/REC-xml-names-19990114/.

Bray, T., Paoli, J., Sperberg-McQueen, C. M. and Maler, E. (2000). *Extensible Markup Language (XML)* 1.0 (Second Edition). W3C Recommendation 6 October 2000. Retrieved December 27, 2000 on the World Wide Web: http://www.w3.org/TR/REC-xml.

Brickley, D. and Guha, R. V. (2000). *Resource Description Framework (RDF) Schema Specification 1.0*. W3C Candidate Recommendation 27 March 2000. Retrieved October 13, 2000 on the World Wide Web: http://www.w3.org/TR/rdf-schema/.

Clark, J. (1999). *XSL Transformations (XSLT) Version 1.0*. W3C Recommendation 16 November 1999. Retrieved October 18, 2000 on the World Wide Web: http://www.w3.org/TR/xslt.

Cochrane Library Issue 3. (2000). *Abstracts of Cochrane Reviews*. Retrieved October 30, 2000 on the World Wide Web: http://www.cochrane.de/cc/cochrane/revabstr/mainindex.htm.

Decker, S., Erdmann, M., Fensel, D. and Studer, R. (1999). Ontobroker: Ontology-based access to distributed and semi-structured information. In Meersman, R. et al. (Ed.), *Semantic Issues in Multimedia Systems. Proceedings of DS-8, 351-369*. Boston: Kluwer Academic Publisher.

Gruber, T. (2000). *What is an Ontology?* Retrieved October 13, 2000 on the World Wide Web: http://www-ksl.stanford.edu/kst/what-is-an-ontology.html.

Grütter, R., Eikemeier, C. and Steurer, J. (2001). Up-scaling a semantic navigation of an evidence-based medical information service on the Internet to data intensive extranets. In *Proceedings of the 2nd Interna-*

tional Workshop on User Interfaces to Data Intensive Systems (UIDIS 2001). Los Alamitos, California, USA: IEEE Computer Society Press.

Heflin, J., Hendler, J. and Luke, S. (1999). *SHOE: A Knowledge Representation Language for Internet Applications*. Technical Report CS-TR-4078, University of Maryland, College Park.

Hendler, J. (2000). *DAML: The DARPA Agent Markup Language Homepage*. Retrieved February 23, 2001 on the World Wide Web: http://www.daml.org.

Hersh, W. (1996). Evidence-based medicine and the Internet. *ACP Journal Club*, July/August, A14-A16.

Horrocks, I., Fensel, D., Broekstra, J., Decker, S., Erdmann, M., Goble, C., Harmelen, F. V., Klein, M., Staab, S. and Studer, R. (2000). *The Ontology Interchange Language OIL: The Grease Between Ontologies*. Tech. Rep. Dep. of Computer Science, Univ. of Manchester, UK/ Vrije Universiteit Amsterdam, NL/AIdministrator, Nederland B.V./AIFB, Univ. of Karlsruhe, DE. Retrieved on the World Wide Web: http://www.cs.vu.nl/~dieter/oil/.

Karp, P. D., Chaudhri, V. K. and Thomere, J. (1999). *XOL: An XML-Based Ontology Exchange Language*. Tech. Rep., Version 0.3.

Lassila, O. and Swick, R. R. (1999). *Resource Description Framework (RDF) Model and Syntax Specification*. W3C Recommendation 22 February 1999. Retrieved July 26, 2000 on the World Wide Web: http://www.w3.org/TR/REC-rdf-syntax/.

Ontopia. (2000). *Solutions for Managing Knowledge and Information*. Retrieved December 27, 2000 on the World Wide Web: http://www.ontopia.net/.

Ovid Technologies. (2000). Evidence-Based Medicine Reviews. Retrieved October 30, 2000 on the World Wide Web: http://www.ovid.com/products/cip/ebmr.cfm.

Sackett, D. L., Rosenberg, W. M. C., Gray, J. A. M., Haynes, R. B. and Richardson, W. S. (1996). Evidence based medicine: what it is and what it isn't. *British Medical Journal*, 312, 71-72.

Schulz, S., Romacker, M. and Hahn, U. (1998). Ein beschreibungslogisches Modell für partitive Hierarchien in medizinischen Wissensbasen. In Greiser, E. and Wischnewsky, M. (Eds.), *Methoden der Medizinischen Informatik, Biometrie und Epidemiologie in der modernen Informationsgesellschaft*. München: MMV Medien & Medizin Verlag, 40-43.

Staab, S., Erdmann, M., Maedche, A. and Decker, S. (2000). *An Extensible Approach for Modeling Ontologies in RDF(S)*. Institut für Angewandte Informatik und Formale Beschreibungsverfahren (AIFB), Universität Karlsruhe. Retrieved July 21, 2000 on the World Wide Web: http://www.aifb.uni-karlsruhe.de/~sst/Research/Publications/onto-rdfs.pdf.

APPENDIX I: SCHEMA OF THE EVIMED ONTOLOGY

(The document has been checked for whether it is well-formed using the XML parser of Microsoft Internet Explorer 5. No further processing was undertaken.)

```
<?xml version='1.0' encoding='ISO-8859-1'?>
<rdf:RDF
   xmlns:rdf="http://www.w3.org/1999/02/22-rdf-syntax-ns#"
   xmlns:rdfs="http://www.w3.org/2000/01/rdf-schema#"
   xmlns:o="http://ontoserver.aifb.uni-karlsruhe.de/schema/
   rdf">
   <rdf:Description ID="Item">
     <rdf:type resource="http://www.w3.org/2000/01/rdf-
     schema#Class"/>
     <rdfs:subClassOf resource="http://www.w3.org/2000/01/rdf-
     schema#Resource"/>
   </rdf:Description>
   <rdf:Description ID="childOf">
     <rdf:type  resource="http://ontoserver.aifb.uni-
     karlsruhe.de/schema/rdf#Irreflexive"/>
     <rdf:type  resource="http://ontoserver.aifb.uni-
     karlsruhe.de/schema/rdf#Asymmetric"/>
     <rdfs:domain rdf:resource="#Item"/>
     <rdfs:range rdf:resource="#Item"/>
     <o:isInverseRelationOf rdf:resource="#parentOf"/>
   </rdf:Description>
   <rdf:Description ID="siblingOf">
     <rdf:type  resource="http://ontoserver.aifb.uni-
     karlsruhe.de/schema/rdf#Irreflexive"/>
     <rdf:type  resource="http://ontoserver.aifb.uni-
     karlsruhe.de/schema/rdf#Symmetric"/>
     <rdf:type  resource="http://ontoserver.aifb.uni-
     karlsruhe.de/schema/rdf#Transitive"/>
     <rdfs:domain rdf:resource="#Item"/>
     <rdfs:range rdf:resource="#Item"/>
   </rdf:Description>
   <rdf:Description ID="parentOf">
     <rdf:type  resource="http://ontoserver.aifb.uni-
     karlsruhe.de/schema/rdf#Irreflexive"/>
     <rdf:type  resource="http://ontoserver.aifb.uni-
     karlsruhe.de/schema/rdf#Asymmetric"/>
     <rdfs:domain rdf:resource="#Item"/>
     <rdfs:range rdf:resource="#Item"/>
   </rdf:Description>
</rdf:RDF>
```

APPENDIX II: RDF SERIALIZATION OF AN EVIMED ONTOLOGY ITEM

(The document has been checked for whether it is well-formed using the XML parser of Microsoft Internet Explorer 5. No further processing was undertaken.)

```
<?xml version='1.0' encoding='ISO-8859-1'?>
<rdf:RDF
  xmlns:rdf="http://www.w3.org/1999/02/22-rdf-syntax-ns#"
  xmlns:rdfs="http://www.w3.org/2000/01/rdf-schema#"
  xmlns:dc="http://purl.org/dc/documents/rec-dces-
  19990702.htm"
  xmlns:odoc="http://ontoserver.aifb.uni-karlsruhe.de/schema/
  ontodoc"
  xmlns:o="http://ontoserver.aifb.uni-karlsruhe.de/schema/rdf"
  xmlns:sontodl="http://.../schema/sontodl">
  <rdf:Description about="">
    <dc:Title>Evimed Ontology</dc:Title>
    <dc:creator>
      <rdf:Bag>
        <rdf:li>Rolf Gruetter</rdf:li>
        <rdf:li>Claus Eikemeier</rdf:li>
        <rdf:li>Johann Steurer</rdf:li>
      </rdf:Bag>
    </dc:creator>
    <dc:date>2000-10-25</dc:date>
    <dc:format>text/xml</dc:format>
    <dc:description>An ontology on evidence-based medi-
    cine.</dc:description>
    <dc:subject>Ontology, Evidence-based medicine</dc:subject>
    <odoc:url>http://...</odoc:url>
    <odoc:version>2.0</odoc:version>
    <odoc:last_modification>2000-10-25</odoc:last_
    modification>
  </rdf:Description>
  <rdf:Description about="http://www.evimed.ch/JournalClub/
  Glossar/RRR.html">
    <rdf:type resource="http://.../schema/sontodl#Item"/>
    <rdfs:label xml:lang="de">Relative Risikoreduktion
    </rdfs:label>
    <rdfs:comment xml:lang="en">This is the definition of the
    labeled concept by Evimed.</rdfs:comment>
    <sontodl:childOf resource="http://www.evimed.ch/
    JournalClub/Glossar/CER.html"/>
    <sontodl:siblingOf resource="http://www.evimed.ch/
    JournalClub/Glossar/ARR.html"/>
    <sontodl:parentOf resource="http://www.evimed.ch/
```

```
      JournalClub/Glossar/NNT.html"/>
   </rdf:Description>
</rdf:RDF>
```

About the Authors

Rolf Grütter is a scientific project manager and lecturer at the Institute for Media and Communications Management, University of St. Gallen, Switzerland. He holds a doctor of veterinary medicine (DVM) from the University of Berne, Switzerland, and a master's degree in business sciences (MSc) from the University of St. Gallen. Rolf has been involved in various projects on knowledge media in healthcare. His research interests focus on knowledge engineering and knowledge media in healthcare, particularly the Semantic Web. Rolf is an active member of the Swiss Society for Medical Informatics (SSMI). He is an editor of the journal *Swiss Medical Informatics*.

***

Lucas M. Bachmann, MD, works as a research fellow at a center for research and dissemination of evidence-based medicine in Zurich, Switzerland, and at the University of Birmingham (United Kingdom). Besides his work in applied diagnostic research, his activities focus on dissemination and implementation of medical information. He teaches evidence-based medicine at the University of Zurich and in continuous medical education programs of the Swiss Medical Association.

Harold Boley, PhD, is a senior researcher at the German Research Center for Artificial Intelligence (DFKI), where he led several government and industrial projects in knowledge representation, compilation, and evolution. Currently, he leads the EU project Clockwork on Web-based knowledge management for collaborative engineering at DFKI. Dr. Boley's focus now is XML-based knowledge markup techniques and RDF-based Semantic Web technologies. He developed the Relational-Functional Markup Language (RFML) and started the Rule Markup Initiative (RuleML). He is also a senior lecturer of computer science and mathematics at Kaiserslautern University, where he conceived AI-oriented XML and RDF courses.

Stefan Decker is working as a PostDoc at Stanford's Infolab together with Prof. Gio Wiederhold in the Scalable Knowledge Composition project on ontology articulations. He has published in the fields of ontologies, information extraction, knowledge representation and reasoning, knowledge management, problem-solving methods and intelligent systems for the Web. He is one of the designers and implementers of the Ontobroker-System. Stefan Decker studied computer science and mathematics at the University of Kaiserslautern and finished his studies with the best possible result in 1995. From 1995-1999 he did his PhD studies at the University of Karlsruhe, where he worked on the Ontobroker project.

Claus Eikemeier is a research assistant at the Institute for Media and Communications Management, University of St. Gallen, Switzerland. He holds a degree in electrical engineering (Berufsakademie Stuttgart, Germany) and in medical informatics (University of Hildesheim, Germany). His diploma thesis was on the subject of Internet-based remote data entry in multicenter clinical trials. His main research focus is in the field of modern Internet technology (XML, Semantic Web, Java), be it the core technology or applied to a specific field—mainly medical. Besides his work at the university, he teaches business informatics at the Polytechnic in St. Gallen and works as a freelancer.

Michael Erdmann gained his master's degree in 1995 from the University of Koblenz (Germany), where he studied computer science and computational linguistics. Since October 1995 he has worked as a junior researcher at the University of Karlsruhe (Germany). He is a member of the Ontobroker project team and his research interests include semantic knowledge modeling with ontologies, and intelligent Web applications based on KR techniques and open Web standards. He is currently engaged in finishing his PhD about the interdependencies between ontologies and XML.

Walter Fierz, MD, studied medicine at the University of Zurich (Switzerland), where he graduated as a medical doctor in 1973. After some clinical experience, he specialized in clinical immunology and has worked in experimental clinical immunology in London (England), Würzburg (Germany) and Zurich. Since 1992 he is head of the clinical immunology section of the Institute for Clinical Microbiology and Immunology (IKMI) in St. Gallen (Switzerland). His year-long interest in medical informatics led in 1997 to a collaboration with the Institute for Media and Communications Management (MCM), aiming at the application of the concepts and techniques of knowledge management in the healthcare sector.

Joachim E. Fischer, MD, MSc, studied medicine in Freiburg, Germany and Dunedin, New Zealand. Residency in anaesthesia and paediatrics. From 1992 to 2001 he was a consultant for neonatology and paediatric intensive care at the University Childrens' Hospital Zurich. He completed his master's of science in clinical epidemiology at the Harvard School of Public Health, Boston from 1997 to 1999. Since 2000 he has worked as a clinical epidemiologist at the Horten-Zentrum, University Hospital Zurich, Switzerland.

Epaminondas Kapetanios obtained his degree from the Department of Statistics and Informatics of the University of Athens, Hellas. He then made postgraduate studies (MSc) at the Technical University (Technische Hochschule) in Karlsruhe, Germany, in the field of database management and knowledge representation systems, before taking up a position as a scientific employee at the Institute of Applied Informatics of the Research Center Karlsruhe–Technology and Environment (FZK). His work addressed the development of a scientific information system concerning atmospheric research. He joined the Global Information Systems group at ETH Zurich, Switzerland, as a research assistant in January 1997 where he finished his PhD in the area of semantics-based querying of data intensive systems. His research interests focus on semantics-based querying and answering, knowledge-based systems and representation as well as the Semantic Web.

Khalid S. Khan, MSc, is a clinician with an interest in promoting knowledge-based practice. Besides his clinical qualifications he has master's degrees in health research methods and medical education. He runs the journal club at his hospital, which is accredited for continuing medical education. He is also the director of the regional postgraduate teaching program for senior clinical trainees in the UK West Midlands region.

Konstantin Knorr studied mathematics and economy at the universities of Mainz and Frankfurt, Germany. Currently, he is a PhD student at the Information and Communication Management Group at the Department of Information Technology, University of Zurich. His research interests encompass security aspects of business processes and formal foundations of computer security. From 1998 till 2000 he worked in the SNF research project MobiMed that focussed on the analysis and implementation of security mechanisms in healthcare environments.

Giordano Lanzola received a degree in electronic engineering in 1988 and a PhD in bioengineering in 1992 from the University of Pavia. He has been involved in both Italian (CNR) and European (AIM) projects concerning artificial intelligence in medicine, in which he mainly focused on the development of advanced knowledge-based systems (KBSs) and on knowledge modeling. More recently his research interests also began to include cooperating software agents. From 1995 on he has been enrolled as a researcher at the Department of Informatics and Systems Science of the University of Pavia, and he is now focusing his research activity on the definition of a multi-agent architecture in healthcare implementing a distributed patient management.

Alexander Maedche is a PhD student at the Institute of Applied Informatics and Formal Description Methods, University of Karlsruhe. In 1999 he received a diploma in industrial engineering, majoring in computer science and operations research, also from the University of Karlsruhe. His diploma thesis on knowledge discovery earned him a best thesis award at the University of Karlsruhe. Alexander Maedche's research interests cover knowledge discovery in data and text, ontology engineering, learning and application of ontologies, and the Semantic Web. Recently, he has started part-time consulting at Ontoprise GmbH in order to apply research on ontologies in commercial practice.

Daniel L. Moody holds a joint position as research fellow in the Department of Information Systems at the University of Melbourne and senior consultant with Simsion Bowles & Associates, an Australian-based information systems consultancy. He is also the current president of the Australian Data Management Association (DAMA). Daniel has held IT management positions in some of Australia's largest organizations and has consulted in strategic information management to a wide range of organizations in both the public and private sectors. He has also published over 40 journal and conference papers in the information systems field. Daniel has extensive consulting experience in the health sector and has developed information management strategies for three of the six state health departments of Australia.

Susanne Röhrig–After graduating in computer science from the University of Erlangen, Susanne Röhrig first worked for the German National Association of Statutory Health Insurance Physicians (KBV) and later in the Department of Data Protection and IT Security of debis Systemhaus. In 1998 she joined the Information and Communication Management group of the Department of Information Technology at the University of Zurich. Since 2000 she has worked as a senior consultant for IT security at SWISSiT Informationstechnik AG, in Solothurn, where she works in the area of network security and security in e-commerce. Her special research interests are in computer security, security management and security of business processes.

Graeme G. Shanks is associate professor in the Department of Information Systems at the University of Melbourne. He holds a PhD in information systems from Monash University. Before becoming an academic, Graeme worked for a number of private and government organizations as a programmer, systems analysts and project leader. His research interests include information quality, implementation and impact of enterprise systems, and conceptual modeling. He has published his research in *Information Systems Journal*, *Journal of Strategic Information Systems*, *Journal of Information Technology*, *Information and Management*, *Requirements Engineering*, *Australian Computer Journal*, and *Australian Journal of Information Systems*.

Allen F. Shaughnessy, PharmD, is a clinical associate professor, Department of Family Medicine, in the Medical College of Pennsylvania/Hahnemann School of Medicine and is director of research and associate director of education for the Harrisburg Family Practice Residency Program, Harrisburg, Pennsylvania, USA. Along with Dr. David Slawson, he has worked on the POEMs project that focuses on methods for physicians to manage medical information. He is a section editor for *The Journal of Family Practice POEMs* and coeditor of *Evidence-Based Practice: POEMs for Primary Care*. He serves as director of HARNET, the Harrisburg Area Research Network.

David C. Slawson, MD, is the B. Lewis Barnett, Jr., Professor of Family Medicine at the University of Virginia Health System in Charlottesville, Virginia. He is the director and founder of the Center for Information Mastery in the Department of Family Medicine and holds a joint appointment as professor in the Department of Health Evaluation Sciences. He also serves as section editor for *The Journal of Family Practice POEMs* and is a member of the editorial board for *The Journal of the American Board of Family Practice,* the *American Family Physician,* and *BMJ USA*. Dr. Slawson is a graduate of the University of Michigan School of Medicine and completed his postdoctoral training in family medicine at the University of Virginia. He is the father of four children, including 3-year-old triplets!

Steffen Staab received a prediploma in computer science from the University of Erlangen-Nuremberg in 1992, an MSE from the Univ. of Pennsylvania in

1994, and the Dr. rer. nat. from the University of Freiburg, Computational Linguistics Lab, in 1998. After a stint doing consulting with the Fraunhofer-Institute for Industrial Engineering, Stuttgart, he joined the University of Karlsruhe, where he is now an assistant professor. Steffen has been working and publishing on computational linguistics, text mining, knowledge management, ontologies, and the Semantic Web. He won a best paper award for a paper on constraint reasoning at ECAI-98. Recently he has cofounded Ontoprise GmbH, a company providing a wide range of technology around ontologies.

Katarina Stanoevska-Slabeva, PhD, studied business administration at the Faculty of Economics of the University "Kiril and Metodij" in Skopje, Macedonia, and accomplished her doctorate at the University of St. Gallen, Switzerland, in 1997 in the field of knowledge management and business process redesign (www.netacademy.org/netacademy/register.nsf/homepages/kstanoevska-slabeva). She is currently working as a lecturer and as a scientific project manager in the research area "media platforms and management" at the Institute for Media and Communications Management at the University of St. Gallen. Her research interests are knowledge media platforms, ontologies, concepts for management of knowledge management platforms, and reference models for component-based knowledge media.

Johann Steurer, MD, is the director of a center for research and dissemination of evidence-based medicine in Zurich. In addition to his clinical qualifications in internal medicine, he has a master's degree in medical education. He teaches evidence-based medicine at the University of Zurich. Besides his research in information technology and transfer he is involved in the development of pre- and postgraduate curricula in medicine.

Hans Rudolf Straub was born in 1954. He finished his studies in psychology and medicine with the final examination in medicine in Berne in 1980. Along with his work as an assistant physician, he began to develop computer programs in 1981 and, since 1986, he entirely works in the field of medical informatics. From 1989 to 1995 he was responsible for the interpretation of medical diagnoses in the first completely "computerized" medical clinic in Switzerland, where he successfully pioneered in the automatic generating of ICD-9 codes out of free medical text. He then continuously improved semantic methods (notation, storage and application of computable knowledge rules) and, together with Dr. H. Mosimann, he founded the company Meditext AG (www.meditext.ch) in 1999.

Stephanie Teufel studied informatics at the Technical University of Berlin and the Swiss Federal Institute of Technology Zurich (ETH), where she received her diploma degree in computer science (MSc) in 1987. Between 1989 and 1990 she was a teaching fellow and lecturer at the University of Wollongong, Australia. Afterwards she was a senior researcher at the Department of Informatics at the University of Zurich, where she received her doctor's degree in 1991 and her habilitation in 1998. From 1999 to 2000 she held a professorship in business informatics at the Department of Informatics at the Carl von Ossietzky Universität Oldenburg, Germany. Since April 2000 she has been with the University of Fribourg, Switzerland, where she has a professorship in management in telecommunication. Furthermore, she is the director of the International Institute of Management in Telecommunications (IIMT) at the University of Fribourg. Her major research interests include: mobile electronic business, information security management, information and communication management, and technology management.

Ulrich Ultes-Nitsche, PhD, is a lecturer at the University of Southampton, United Kingdom. He studied computer science and physics at the University of Frankfurt. Before joining Southampton, he was a junior scientist at GMD Darmstadt, a visiting researcher at the University of Liège, and a research assistant at the University of Zurich. He received in 1995 a DAAD HSP II/AUFE fellowship and in 1997 the Best Student Paper award of the 16th Annual ACM Symposium on Principles of Distributed Computing (PODC •97), Santa Barbara, USA. His research interests comprise security, formal methods and automated verification. His Web page is at http://www.ecs.soton.ac.uk/~uun.

Markus Wagner is research fellow at the Institute for Medical Statistics and Documentation of the University of Mainz Medical School. He is involved in projects of the German Health Research Network and the Coordination Centre for Clinical Trials at Mainz. His scientific interests include medical decision support systems, knowledge representation, inference algorithms and clinical data management. In 1998 he graduated from the University of Koblenz, Computer Science Faculty. His diploma thesis was about pharmaceutical information systems and the automatic detection of serious medication errors. The thesis received an award by the German Medical Informatics Association (GMDS) in 1999.

Index